AF879737

A Beautiful Mind and Soul

A Catholic Scientist Explains

Dr. Gerard Verschuuren

En Route Books and Media, LLC
Saint Louis, MO

ENROUTE
Make the time

En Route Books and Media, LLC
5705 Rhodes Avenue
St. Louis, MO 63109

Cover credit: Sebastian Mahfood

Copyright © 2022 Gerard Verschuuren

ISBN-13: 978-1-956715-25-5
Library of Congress Control Number: 2022930995

No part of this book may be reproduced, stored in a retrieval system, or transmitted in any form, or by any means, electronic, mechanical, photocopying, or otherwise, without the prior written permission of the authors.

G.K. Chesterton

"the more we really look at man as an animal,
the less he will look like one."

Table of Contents

Preface

Science never talks about mind and soul—and if it does it has no nice words to say about them. So how can a Catholic scientist speak about "a beautiful mind and soul"? Well, he does so as a Catholic who knows that science has nothing to say about mind and soul, but also that science has nothing to say against it.

What I want to show in this book, as a Catholic scientist, is that humans do have a beautiful mind and soul, period. But to show how true this is we need to clear many hurdles first. Many of these barricades were erected by science. Or perhaps I should say, they were not put up by science but by scientists who placed these obstacles on our way "in the name of science." Science properly understood does not do this and cannot do this, but its scientists have often tried, making us believe that it was science causing these problems.

I want to achieve my goal with a rather unusual format by using a lively, conversational dialog: an exchange between a skeptical scientist (S) and a religious scientist (R). The skeptical scientist interrogates the religious scientist with very penetrating questions. And the answers given will hopefully and ultimately clear the way to become appreciative of the beautiful mind and soul that all of us were given. In other words, it is not a debate between two scientists about scientific matters. Rather it is a debate about "life and death" issues seen from the perspective of two different kinds of expertise.

During this dialog, we will try to remove the obstacles that have been put on our way by supposedly science-minded, skeptical obstructionists. We will do this by using the following timeline: first we need to clear the path from obstacles put up by those who see a conflict between science and religion; then we need to explain why science cannot block the way to accept the existence of soul and mind; next, we need to clear all the obstacles, one after the other, which we encounter on the way to get to the beautiful soul and mind of human beings; after that, we need to make clear what it is that the existence of soul and mind, of rationality and morality, adds to human life; and finally, we should address whether soul and mind can survive after death.

In doing all of this, we will find out that there is still room left for religion next to science, that there is still room left for the soul in addition to the body, that there is still room left for the mind along with the brain, that there is still room left for a moral code over personal likes and dislikes, and that there is still room left for the soul transiting from earth to heaven.

The title of this book may remind some readers of the 1997 biography and 2001 movie called *A Beautiful Mind*, which eternalized the life story of the 1994 Nobel Prize laureate John Forbes Nash Jr., played in the movie by actor Russell Crowe. However, John Nash's life also had a darker side. In 1959, he was diagnosed with paranoid schizophrenia. During the 1960s, he went through several periods of psychiatric in-patient care during his battle with the paranoid delusions caused by the disease.

Why did the writer of the biography and the makers of the film decide on the title *A Beautiful Mind*? They could have spoken of "a beautiful brain," as an ingenious intellect is usually connected with the brain. Besides, psychiatric diseases are typically associated with genes and neurons—again, a brain issue not a mind issue. Yet, they decided to refer in their title to the mind instead of the brain. I do not really know what their motives were, but I do know why I wanted to make this title part of the title to this book. The thesis of my book is that the mind is much more than the brain, which explains why the mind can be called "beautiful," but not the brain. The mind made it possible for Nash to gradually get back on track in his life and to be the person he really was. That is the power of the mind, even when the brain is not quite functioning properly. The brain is just a biological part of the body, but the mind comes with a beautiful soul. So that's how the title of this book came about.

I promise you it will be an enlightening tour through the pivotal questions raised by our human minds and souls, which were created in God's image. I trust and pray that on your way through the dialog in this book, you will come across many valuable answers and insights.

CHAPTER 1

When Science Wants the Last Word

According to many nowadays, religion seems to be on its way out, being replaced by science. The battle between the two looks lost if we go by opinion polls. The reason is probably that science has such a great track record, and it has taken over questions that used to be considered religious. Besides, it has the mass media on its side. Famous scientists and atheists have inundated the market with their anti-religious propaganda. But is it really a lost battle for religion?

I am a religious scientist (R) who would like to ask my opponent, a skeptical scientist (S), what it is that makes some scientists think that the battle is over, with science as the winner. In response, the skeptical scientist probably wants to question my credentials.

The Conflict that Wasn't

S: Do you really want to announce yourself as a "religious scientist? How can you call yourself a scientist if you want to be religious at the same time? Science and religion do not go together. One of the two must surrender, for they contradict each other. Obviously, I think it is religion that is on its way out.

R: I do not share your view that science and religion are in a conflict with each other. When I say I that I come from God, I am not denying that I came here through my parents, and that they in turn came here from their parents. There is nothing wrong with saying that children come from their parents, and yet at the same time, that children come from God—no contradiction here. On the one hand, without God, there would be no children. On the other hand, without parents, there would be no children either. I am not denying either side of the equation. Your idea of a conflict is a myth; there is no conflict here. If the existence of God is true, and if certain scientific conclusions about the natural world are also true, then they cannot contradict each other—both sides are true.

The role of my parents can be studied by science, more in particular by genetics, but that does not mean I can no longer maintain that I came here from God. The former part comes from science, the latter part from religion. Yet, both can be true and are in fact true at the same time. There is no contradiction here, no conflict, no battle, no victory for one of the two. Neither one can be forced to leave. Only when I would say I came here either from my parents or from the stork, one of the two cannot be true. But that is not so about my parents versus God.

An interesting case can be made with the Big Bang theory. The momentous question that stirred scientists in the early 1960s was whether the universe had a beginning in time. Many of them at the time were opposed to this idea, because a first moment of the universe seemed to suggest a "point of creation," which is an idea typically linked to religion. Even Albert Einstein himself was afflicted with this prejudice. When he used his new theory to construct the gravitational equations governing the universe, they suggested that a universe dominated by gravity would either expand or contract. This idea was so philosophically "repugnant" (his word) to him that he added a constant to the equations to balance them out, so they would describe an eternal universe. It was a "fudge factor" to block gravitational implosion. He later called this his "biggest blunder." The consequence of his blunder was that he failed to predict the cosmic expansion that Edwin Hubble would measure in 1929.

Then, in 1931, Fr. Georges Lemaître, a Catholic priest and physicist from Belgium, suggested that the evident expansion of the universe, if projected back in time, meant that the further in the past, the smaller the universe was, until at some finite time in the past, all the mass of the universe was concentrated in a single point. What came out of this scientific discussion is that the universe was no longer eternal but must have had a beginning—something religion had been saying for centuries. Amazingly, a key role in this discussion was played by a priest who found no conflict between his ancient faith and the revolutionary findings of the newest science. He was later lauded by the Nobel Laureate Paul Dirac with the words, "we must rate Lemaître's cosmology of the highest caliber."

I think it is very telling that Arno Penzias, who had received the Nobel Prize in Physics together with Robert Wilson for their observation of the "cosmic microwave background," would later write of the origin of the universe, "The best data we have are exactly what I would have predicted, had I nothing to go on but the five books of Moses, the Psalms, the Bible as a whole." The point I am trying to make, as did Penzias, is that religion and science can very well be in harmony with each other, without any conflict. Science and religion are the two "windows" that let humans look at the world they live in. Yet there is only one world, and it is the same world these two windows try to reveal to us. Closing off either window, as the conflict theory wants us to do, would make us partially blind.

S: Really? I actually think religion had gone already a long time before the above discussion took place! More and more people, especially in science, had already left religion behind, and most of them will tell you that they have much more confidence in science than in religion.

R: It is funny you say this, for there has always been a strong bond between science and religion. It was the Franciscan Friar Roger Bacon in the thirteenth century who, together with Bishop Robert Grosseteste and the Dominican

Friar St. Albertus the Great, clearly articulated the "modern" conception of (natural) science. In doing so, they were the forerunners of the era in which the so-called revolutionary science came to life (1300–1650).

Very soon in the history of science we find many religious believers who connected their scientific work with God's work in nature. It is striking how the greatest scientists of all times, men like Nicolaus Copernicus, Johannes Kepler, Isaac Newton, and Gottfried Leibniz, were inspired by their deep religious beliefs. These scientists explicitly claimed they could achieve in science what they did because of their religion. Johannes Kepler, for instance, once beautifully summarized this conviction, "The chief aim of all investigations of the external world should be to discover the rational order and harmony which has been imposed on it by God."

Well, this is not only true of the "old" days. Repeatedly, we see scientists arise at important moments in the history of science calling themselves religious, or even Catholic. Recent giants in science that come to mind are Fr. Gregor Mendel in genetics, Louis Pasteur in microbiology, André Marie Ampere in physics, Fr. George Lemaître in astronomy, Max Planck in quantum physics, and so many others. This entitled Max Planck to declare that "the greatest naturalists of all times, men like Kepler, Newton, Leibniz, were inspired by profound religiosity." No conflict here!

Yet, as you mentioned, there is this deep-seated prejudice that science and religion cannot live in harmony. When renowned geneticist Francis Collins, the former Director of the famous *Human Genome Project*, was going to become nominated to be the new director of the *National Institutes of Health*, several scientists and pundits publicly questioned whether the nominee's devout religious faith should disqualify him from the position. Nevertheless, Collins was unanimously confirmed by the U.S. Senate on Aug. 7, 2009, but the controversy over his nomination reflects a growing rift between those who believe religion and science each examine legitimate but different realms of knowledge and those who see science as the only true way of understanding the universe.

It was this very Francis Collins who once wrote "that the principles of faith are, in fact, complementary with the principles of science." He is a current example of a dedicated scientist and a dedicated Christian, who testified against the one-sided views on both sides of a seemingly conflict between science and religion. He testifies that we do not need to turn our backs on science, nor do we need to turn our backs on religion. Collins again, "The God of the Bible is also the God of the genome. He can be worshipped in the cathedral or in the laboratory. His creation is majestic… [and] cannot be at war with itself." That's why we can say, at the same time, that we come from God and from our parents.

A Forgotten History

S: That may be true, but it is very well possible that all the religious scientists you mention could have achieved the same results without religion. Their achievements in science did not come from religion but were only possible in spite of their religion.

R: I am not so sure about that. I could in fact go one step further: science was in fact born in the cradle of religion. That means we would have no science without religion from which it came forth. My claim is that science is just impossible without the impact of religion, especially Christianity.

There must be a reason why there can be no science without the Judeo-Christian tradition. The reason for this connection is simple: the Bible teaches us that the universe was *created*. For science to arise from a religious cradle, the most important condition is a deep and strong belief in a Creator God. It is this very Judeo-Christian concept of a Creator God that makes science possible. Belief in a Creator God entails that nature is not a divine but a created entity; nature is not divine in itself, only its Maker is. It is this belief that opens the door for scientific exploration. Nature, the world, the universe, they all were created by the rational intellect of a Creator capable of being rationally interrogated. It's only in such a cradle that science could be born and begin its journey of scientific exploration.

The Book of Wisdom (11:20) says about God, "You have arranged all things by measure and number and weight." If that is true—and it is in my opinion—then the only way to find out what the Creator has actually done is to go out, look, and measure so as to interrogate nature by observation and experiment—which is the beginning of scientific exploration and investigation. From that point on, science is on its way, but could not have emerged sooner.

S: Didn't others start science as well without being religious, let alone Catholic? I don't think religious people were the only ones to bring about science. They were rather the exception. We find also scientists in history who were outside the Judeo-Christian tradition.

R: I don't think so. As a matter of fact, it can be shown that peoples not familiar with the Judeo-Christian tradition failed to nurse science. Cultures like ancient Greece, Egypt, and China thought nature was ruled by whimsical deities. In such an environment, science didn't have a chance and wouldn't have made much sense. Even Greek philosophers such as Aristotle and Pythagoras actually invented their own god, bound by the rules they had made up on their own.

It could be argued that a tendency toward a different conception of God is part of what distinguishes other religions from the Judeo-Christian religion—which might explain why natural science improved in the West and weakened, or even lacked, within the rest of the world. Because the Judeo-Christian God is a reliable God—not confined inside the Aristotelian box, not capricious like the Olympians in ancient Greece, and not entirely beyond human comprehension as in Islam—the world depends on the laws that God has laid down in creation.

This explains why science was born in the cradle of the Catholic Church and could not have been born anywhere else—not in China (with its sophisticated society), not in India (with its philosophical schools), not in Arabia (with its advanced mathematics), not in Japan (with its dedicated craftsmen and technologies), but on Judeo-Christian soil. Only in Europe did alchemy develop into chemistry and did astrology lead to astronomy. The physicist Fr. Stanley L. Jaki used the phrase "stillbirths of science" in reference to the ancient cultures of Egypt, China, India, Babylon, Greece, and Arabia.

In other words, the Judeo-Christian tradition was needed for science to arise. In Jaki's own words again, "Within the biblical world view it was ultimately possible to assume that the heavens and the earth are ruled by the same laws." The nuclear physicist J. Robert Oppenheimer, not a Christian himself, agreed with this when he said, "Christianity was needed to give birth to modern science." In other words, there is no conflict between science and religion, but rather a deep-rooted harmony. Therefore, we must conclude that science and religion not only are able to live together, but also must live together.

I agree with you, though, that the idea of a conflict between science and religion is still very much alive and is even getting more traction lately. Conflict stirs the media, whereas harmony is considered boring. However, the so-called conflict theory is not the outcome of scientific research but rather the assumption of what some people like to believe about science. The conflict theory was born and conceived during the age of the Enlightenment in the 17th century. It is not a "theory," but at best a "hypothesis"—and a poor one at that. It assumes a make-believe conflict, often backed up by the infamous Galileo affair. Believe it or not, there he comes again: Galileo!

However, a cloud of myths and lies obscures the real facts about the Galileo case: he did not discover the earth is round; he did not invent the telescope; he was not the first to advance heliocentrism; he could not even prove heliocentrism; he was not silenced by anti-scientists; he was not tortured by the Inquisition; he was not incarcerated by the Inquisition; he did not mutter, "And yet the earth moves." In other words, the conflict theory is based on myths—it has no foot to stand on. It is merely a wish or desire in the minds of some, but certainly not in the minds of the early scientists I mentioned earlier.

My point is that the Galileo affair has been blown up to unfair and unjustified proportions. The nuclear physicist

Stephen Barr testifies to this: "Ask a Catholic audience whose name they associate with the Catholic Church and science. 'Galileo!' they shout." I admit, there were mistakes made on both sides of the conflict at the time, but there was no conflict here between science and religion. Interestingly enough, St. Thomas Aquinas had thought about this problem as early as four centuries before Galileo. He astutely noticed that the appearance of the visible motions of the celestial bodies "are produced either by the motion of the object seen or by the motion of the observer… it makes no difference which one is moving." In other words, the sun could be moving, or we on earth could be moving. How profound are these words, coming from the mouth of a religious scholar whom many scientists would call outdated nowadays.

The point I am making here again is that the conflict theory definitely is a myth to make the sweeping claim that religion, and the Catholic Church in particular, has always obstructed science from doing its work. It is obviously a myth promoted in media and academia.

A Division of the Estate

S: Well, that's how it may have started, but science is grown-up now and has become fully independent. The earliest scientists were merely pioneers who hadn't gained their independence yet. Finally, they saw the light, thus making it possible to sever the ties with religion.

R: You're right in the sense that science would gradually articulate more clearly what its actual territory is. When scientific societies began to blossom in 1650, the Fellows of the *Royal Society of London*, for instance, received from King Charles II the privilege of enjoying intelligence and knowledge, but with the following important stipulation "provided in matters of things philosophical, mathematical, and mechanical." All the rest was given away for others to handle, as long as their own members, the future scientists, were permitted to "know" what can be known by counting and measuring.

That's how the "division of the estate" was executed. By accepting this separation, science bought its own territory, but certainly at the expense of inclusiveness. The rest remains part of a different territory and is therefore beyond the reach of science, lying outside its scope. These scientists realized very clearly that religion would always be off-limits to them. But later scientists seem to have forgotten this. As long as they keep forgetting this, they keep the conflict myth going without having any reason to do so. It's a prejudice they foster—and prejudices can stay alive for a long time. As Albert Einstein once mentioned, "It is easier to split atoms than prejudices."

The conflict myth made headlines again when, on Christmas Eve 1968, three astronauts in the first manned spacecraft to orbit the moon read from their capsule to the world the first ten verses of Genesis 1. Some atheist filed suit against NASA for permitting this Christmas Eve reading of the Bible—in an effort to ban public prayer from space. It would not be the last skirmish caused by the conflict myth.

In defiance of this myth, the late Harvard biologist Stephen Jay Gould said it quite clearly: "Science can only work with naturalistic explanations; it can neither affirm nor deny other types of actors (like God) in other spheres (the moral realm, for example)." The Austrian physicist and Nobel Laureate Erwin Schrödinger once said something similar about science, "It knows nothing of beautiful and ugly, good or bad, God and eternity." The three astronauts basically testified to this in what they did.

S: I keep stressing, though, that the scientific territory has grown extensively over the centuries. What science has gained since was gained at the cost of religion, thus leaving fewer and fewer issues for religion. Scientific expansion means religious withdrawal—so religion must be on its way out whenever science advances.

R: I disagree with you that the expansion of science has been at the cost of religion. Let me explain where the misconception lies. Science does not gain territory at all when it makes new discoveries, but it just learns more and more details about its own fixed and demarcated territory—which is the domain of all that can be dissected, counted, measured, and quantified. The rest is not part of its territory but was given away for other "authorities" to handle, as we discussed already.

If you want to picture this process with an analogy from sports, perhaps the following images help. Somehow, you are comparing the "confrontation" between science and religion with what happens on a *football* field – where advancements on one side make the other side retract. But I would say the dialogue between science and religion takes place on a field that is more like a *tennis* field, where each party should stay on its own side to do its "performances."

In other words, rather than expanding their territory, scientists have become more "specialized" in specific fields and branches of science. One might even go as far as the late Nobel Laureate and biologist Konrad Lorenz who famously said that a scientist "knows more and more about less and less and finally knows everything about nothing." I know that is extreme, but there is something to it. It is in fact scientific arrogance to claim universal validity for local successes in science—in defiance of the fact that the astonishing successes of science have not been gained by answering every kind of question, but precisely by refusing to do so.

I conclude from this that science and religion are not competitors; they are rather "neighbors," each one with their own domain, perspective, and authority. Good fences make good neighbors, as the saying goes. It is in their respective domains that science and religion function best. If we do not honor the separation between the two, science may become a pseudo-religion, and religion may become a semi-science. Instead, we should keep them separate. If there is any conflict between science and religion, it is largely a conflict between some men and women of the science community and some men and women of religious communities, rather than between science itself and religion itself.

The Ties that Cannot Be Severed

S: Perhaps you are right that science did not take over what used to belong to religion, but it did make religion practically redundant. Science can do well, or even better, without religion. It has come up with more and more answers to questions that were once considered to be religious.

R: I keep maintaining that scientists can never really sever their original ties with religion. If they do, they undermine their own activities. I would even like to put it this way: scientists keep living off religious capital, no matter how "independent" they think they have become. What do I mean by that?

Well, for them to do their work in science, scientists need to uphold certain *assumptions* that science cannot provide itself—yet are needed for science to work. Without those assumptions, science has no base to rest on. Science cannot take credit for its own starting point. Think of the following. Without the assumption that there is an objectively real world, science would not be possible. Without the assumption that there is order in nature, science would not be possible. Without the assumption that like causes have like effects, science would not be possible. Without the assumption that the world is comprehensible, science would not be possible. And yet, none of these assumptions can be proven by science.

In short, all scientists assume—and must assume, even though they do not always acknowledge it—that the universe is real, orderly, causality-based, and comprehensible. In other words, science is something you need to "believe" in before you can practice it. Science cannot even begin without those assumptions. What science cannot do is proving its own assumptions, yet they are needed for science to do what it can do. In other words, what science can do is only possible if it realizes what it *cannot* do—testing and proving its own assumptions, to begin with.

Where then do those assumptions come from, you might ask. Apparently, not from science itself, for that would amount to circular reasoning—assuming ahead of time what one attempts to prove. Their source must come from somewhere else. Well, I may have a surprise for you: that source is religion. Scientists must borrow from the Judeo-Christian faith what they themselves cannot provide. Once they consciously sever those ties with religion, science can no longer get off the ground. That would be like cutting off the very roots from which science did spring up. Yet, some scientists prefer to just keep silent about these assumptions underlying science, so they think you neither need to admit you hold them nor need to explain why. Keeping silent would probably make one's work as a scientist much easier. But how long can one keep doing that and remain silent?

I would like to turn roles now and ask you a few pertinent questions: How could nature be intelligible if it were not created by an intelligent Creator? How could there be order in this world if there were no orderly Creator? How could there be laws of nature if there were no rational Lawgiver? Yet, there are laws of nature that science is in search of, which calls for and points to a Lawgiver.

S: Your stock argument seems to be that there can be no laws of nature without a lawgiver. However, scientists use the phrase "laws of nature" less and less. Instead, they see those "laws" as pure descriptions of cause-and-effect relationships.

R: I think by saying this, you are turning the situation upside down. The laws of nature did not *gain* a new status on their own, but some scientists decided to *give* them a new status by changing their own terminology. Their main reason for doing so was that the term "law" almost automatically reminded them of a "lawgiver," which they didn't like, because the phrase "laws of nature" seemed to suggest they were implemented by a higher authority, a Divine Lawgiver. So, they thought they could get rid of this Lawgiver by a simple terminological shift. They developed some form of law-phobia and decided to disconnect their laws from a higher authority.

Did this make the problem go away? Of course not. The reason is simple: even if we declare the laws of nature as pure descriptions of cause-and-effect relationships, we are still left with the question as to where the regularity behind these relationships comes from. We won't wake up tomorrow to find water flowing from a lower point to a higher point, or heat flowing from cold to hot. That would be a violation of the "law and order" of this universe. Even cause-and-effect relationships are part of an orderly universe—which shifts the question to where this order comes from. Interestingly enough, the word "cosmos" is the Greek word for "order."

No matter how we call the laws of nature, or prefer to call them, the key question remains the same: where do the laws of nature—or the "law and order" of the universe—come from? They cannot come from themselves, nor can they generate themselves. The laws of nature which govern an existing universe can never generate that universe itself and bring it into existence, as some scientists have claimed, for that creates a logical conflict. For something to create itself, it would have to exist before it came into existence—which is logically and philosophically impossible. Neither can the universe generate its own laws of nature, for that would still require other laws of nature. In other words, the laws of nature cannot bring themselves into existence and cannot explain themselves. Hence, they must come from somewhere else. I think the only reasonable answer is this: they come from a Lawgiver who created the world and implemented laws of nature.

To think that science has shown us there is no God and no Lawgiver is in fact an argument that starts at the wrong end. Fr. Francis Canavan gave us a good reflection on this issue when he wrote about people who reject the existence of a Creator: "When we have fully peeled the onion of the godless universe, we find at its center nothing at all.... Men do not stop believing in God because they have discovered that the world is meaningless. Rather, they conclude that the world is meaningless because they have stopped believing in God." In contrast, we should reverse the reasoning: if nature is an image or reflection of God, its Creator, then to study nature is an indirect way of studying God. Galileo could not have said it better when he refused "to believe that the same God who has endowed us with sense, reason, and intellect has intended us to forgo their use."

All of the above is a rather long answer to your original short question as to whether a scientist can be religious. Scientists can certainly be religious. One might even say that they *must* be religious in order to do their work in a coherent way. I know that is quite a claim—hard to accept for some—but not an unreasonable one on further investigation, so we found out. Some people may think that science has the last word in this discussion, but one might as well say that religion ultimately has the last word.

CHAPTER 2

The Side Effects of Science

Many people have a rather simple view of what science does. They think science is merely a matter of observation, experimentation, measurement, and quantification. This makes it look like science is an objective, neutral, fact-based enterprise. That may be true, but there is much more to it. A closer look at science, with the eyes of a *philosopher*, reveals what is missing in this picture. Science also comes with what one might call unknown or unexpected side effects.

S: When you talk about science, or about religion for that matter, you always confuse me by interjecting lots of philosophy. What does philosophy have to do with this?

R: For a starter, there is probably not one person who does not philosophize at some point in life. Pope John Paul II went even as far as saying, "the human being is by nature a philosopher." That must also apply to scientists. Of course, this does not automatically result in high-end philosophy, but it is a starter. It may never lead to any kind of real, serious, or solid philosophy, though. We tend to use the word "philosophy" for many things, often as a shorthand for any kind of worldview. When I talk about philosophy in this dialog with you, however, I am referring to a more systematic system of thoughts. There are many kinds of philosophic systems. A broad distinction among them would be the separation between sound philosophies and shaky philosophies. A sound philosophy, such as the philosophy of Aristotle or St. Thomas Aquinas, can be the "glue" that holds science and religion together, whereas shaky philosophies, such as materialism and relativism, place obstacles in their way.

Let me be brief on relativism (I will get back to it later). It claims that there is no absolute truth—there are only opinions about what is true, depending on one's point of view. This means, what is true for you may be false for me. It promotes a sort of democratic ideal in matters of truth and knowledge. What is wrong with that idea? Well, when we say that the earth is not flat, we make a statement that is either true or false. Holding the opposite opinion that the earth is flat does not make the earth flat, even if you would like it to be flat. Truth is truth, even if you do not accept it; and untruth is untruth, even if you claim it. This makes relativism a shaky kind of philosophy, an "argument stopper"—the end of any discussion.

In other words, truth is not something we create or invent, but something that we try to "capture" in the real world. That's what all science is about—the truth. Besides, the problem with a claim like "There is no such a thing as objective truth" is that it claims to be objectively true. We cannot make such a sweeping assertion without also asserting, implicitly, that there is such a thing as objective truth after all. That takes us back to where we started. It is a kind of philosophy that I would not recommend to anyone.

S: Let me agree with you for now. However, when you mention the side effects of science, what do you mean by "side effects"? Are you telling me that science, like medications, can have dangerous side effects, too?

R: Perhaps I should not say that science itself has side effects, unless they are taken as neutral or positive side effects. The dangerous side effects, on the other hand, come from the scientists who draw conclusions from their research that are unwarranted by what they have discovered. Scientists are indeed experts in their specific fields, but this makes them sometimes claim they are also authorities in areas beyond their scientific expertise. Philosophy is arguably one of them.

It always strikes me how some scientists, with several Nobel Laureates among them, think that their authority and expertise in a tiny segment of science entitles them to proclaim authoritative statements about everything else outside that small segment. They use the scientific expertise in their field to make us think they must be experts in everything else as well. They often use the trump card of "science says so," but it is not always clear "what science says." Science is not a monolithic enterprise with only one set of undisputable verdicts. It is not a matter of "what science says," but of "what some scientist says." Some scientists go as far as using their self-proclaimed authority to churn out, with the speed of light, books that promote opinions and views that go far beyond their field of expertise. Scientists with this view either are bad philosophers or they assume that everyone else is. Perhaps Albert Einstein was right after all when he said: "It has often been said, and certainly not without justification, that the man of science is a poor philosopher."

Do not take this as a critique of science and its methodology. It is a critique of the scientists who abuse their authority to make philosophical claims that go beyond their scientific research, making them think they can use the technique of raising question-begging philosophical objections disguised as scientific objections. Yet, the results are usually very disappointing. Again, that's not science's fault—we should have never expected from science what it cannot deliver. My point is that science is not to blame for these side effects—scientists are. Let's find out what this has led to.

The Dismantling Approach of Science

S: I don't know what you mean when you characterize science as a dismantling activity? I would say that this feature puts science in a dubious position. Dismantling sounds to me as a demolishing activity. I don't think that is fair to say about science.

R: Nevertheless, I maintain that science is typically based on dismantling its object of investigation. It "peels off layers," so to speak, until it reaches deeper layers or lower levels of investigation, ultimately going down to the lowest level or deepest layer, subatomic particles or so. Seen this way, the dismantling approach of science is not really a demolishing activity. Scientific research has sometimes been likened to a "warfare against the unknown." This image suggests an important tactic. The procedure most likely to lead to an advance is that of concentrating one's forces on a very restricted sector which is chosen because the "enemy" is believed to be weakest there. Dismantling things may in fact be a smart strategy of concentrating on a restricted area of investigation.

Let me explain the idea of science as a dismantling activity with a simple analogy. When we study how a machine works, we can do so at several levels. At the "top" level, the level of the machine itself, we can explain what the machine can do for us. This is basically what a user's manual tries to achieve. But we can go one level deeper or lower by describing the individual components of the machinery and explaining how each of them works. The regular user will probably not be interested in that level. Engineers are; they will typically go even one more level down to understand how the components can do what they do. Is that "deep" or "low" enough? It depends. Some may want to know what happens inside those components at a physical or even quantum level.

What I try to show with this crude example is that we can analyze things at different levels. Depending on your need, you can "peel off" certain layers until you get to the layer where you want to be. For biologists, that might be the level of cells; for biochemists, the level of molecules; for physicists, the level of atoms or subatomic particles. Based on this, I would even go as far as claiming that reality itself is a multi-layered phenomenon. That's why dismantling works.

S: Where do your "layers" come from? What are they based on? It seems to me that you introduce the idea of a multi-layered reality so as to counteract any monolithic claims of science. That sounds rather self-serving to me.

R: It is not unusual at all to distinguish several layers in reality—even so, or specifically so, in science. It is rather common among scientists—especially those in the life sciences, the behavioral sciences, and the social sciences—to distinguish various levels of organization in a so-called hierarchy of entities, also called an *organizational hierarchy*. Usually, this hierarchy is seen by its users as something based on reality, not just on a mere arbitrary scheme of classification. Each level in the hierarchy is seen as composed of "building blocks" from "lower" levels, but the level itself is also "building material" for "higher" levels.

A rather detailed and extensive division from "low" to "high" levels could go this way: atoms | molecules | cells | tissues | organs | organisms | populations. This list can be read as follows: atoms are parts of molecules, molecules are parts of cells, cells are parts of tissues, tissues are parts of organs, organs are parts of organisms, and organisms are parts of populations. For good reason, this construction is typically called a "hierarchy"—not so much because the science of a "higher" level is really "on top of" the previous one, but because entities at each level can be understood in either direction: in terms of their components (that is, at a "lower level") or in terms of what they themselves compose (at a "higher level"). Therefore, atoms are parts of molecules, but also the other way around, molecules are composed of atoms.

Consequently, scientists nowadays consider themselves specialized not so much in the *object* of their research as in the *level* of their research. By this they mean that they prefer to work at a certain level inside a much larger hierarchy. They focus on what is specific for a certain level and thus abstract from other levels. Each level spans a specific range of phenomena. Biology, for instance, is less general in scope than chemistry; and chemistry is less general than physics—whatever that actually means. Physical and chemical properties are present in anything there is, but biological properties are said to be present only in systems specifically recognized as biological objects.

S: I admit, the way you describe this approach seems very legitimate to me. I don't see anything wrong with your analysis. It has in fact been very productive and successful for science. So how then can you describe this approach as a "dismantling" activity?

R: I agree with you, there is nothing wrong with going down to lower levels of reality—ultimately to the lowest level of atoms and subatomic particles, that is. This approach may in fact be sound scientific methodology which has provided science many successful inroads. It is often called *reduction*. But what should indeed be considered wrong and unwarranted is claiming that this lowest level is the only level that counts and makes for the one and only explanation of all the other levels above it, which would be *reductionism*—a philosophical worldview, not a scientific conclusion, let alone a scientific technique. It is a side effect of science that some scientists like to promote as if it were scientific. Taken to its extreme, reductionism claims that the ultimate goal of science is to explain everything—yes, everything!—in terms of its smallest pieces at the lowest level. Whereas reduction is certainly a successful scientific technique, reductionism is an unwarranted worldview, an ideology, or a doctrine.

Why unwarranted, you may ask. What reductionism boils down to is speaking of everything in this world in terms of "nothing but"—that's why this ideology is often called "nothing-but-ery" or "nothing-but-ism." It is a kind of myth that you will hear very often when scientists leave their own territory, their own little area of expertise, and then blow it up to immense proportions. Just like some movie stars, so do some scientists abuse their fame to also ventilate their worldviews—or lack thereof—convinced as they are that their professional expertise guarantees the quality of any other beliefs they might have. I would rather warn you to never let a plumber fix the electrical wiring in your house! I would advise plumbers and electricians to stay with their trade.

Examples of reductionism are plentiful and have occurred at all levels of the organizational hierarchy. Here are the most telling ones, each one dealing about ourselves as human beings:

1. At the level of the universe, we as human beings have been degraded to *nothing but* "stardust," in the words of the astronomist Carl Sagan. The message is clear: away with ideas of superiority. At one time the earth was put in the center of the universe (geocentrism), but later turned out to be just one of the million planets around us—and an insignificant one besides. As a consequence, no longer are human beings the supreme creatures of creation, but merely the insignificant inhabitants of an insignificant planet making them feel

lost in this universe with its enormous size and its immense age.

Is this really science, or should we call it science-fiction? To say the least, it is reductionism. It is a creed directed against religious people who tend to claim that the earth was made for us as the crowning glory of creation. The latter claim is not at odds with reduction, but with reductionism it is. Besides, this form of reductionism forgets that all the knowledge we have about the universe is only possible thanks to the minds of human beings—of those called "stardust" by this ideology.

2. At the level of society, we have been degraded to *nothing but* a number, a statistic, a member of a colony. This is the level where the social sciences seem to feel at home. Reduction to this level is, again, a legitimate technique, but when transformed into reductionism, it becomes a deceiving ideology. It turns each human being into a mere means for the sole purpose of advancing society—in the same way as termites, ants, and bees only exist to keep their colony alive.

However, at the very moment we glorify the "colony" over its "members," individuals lose all their rights and become mere slaves of the totality—which leads to various kinds of pure totalitarianism. The horrors placed on human beings by totalitarian regimes—in the labor camps run by Nazis, Soviets, and Maoists, for instance—tell us how ugly this ideology can get. It makes a society of human beings look like a colony of termites or bees. As Pope Pius XI put it in a nutshell, "Society is for man, not vice versa." When Pope John Paul II issued *Centesimus Annus* in 1991, he diagnosed the root problem as follows: "Socialism considers the individual person simply as an element, a molecule within the social organism, so that the good of the individual is completely subordinated to the functioning of the socio-economic mechanism." Those who think this way do not "follow science," but have fallen victim to an ideology.

3. At the level of species or populations, we are degraded to *nothing but* overvalued animals. We breed, bleed, feed, and excrete just like all animals do. This is the level where biology feels at home. Reduction to this level is a legitimate technique, but when transformed into reductionism it becomes a dangerous ideology. It obscures the fact that we have faculties no other animals have. Some animals may out-see us, out-smell us, out-hear us, out-run us, out-swim us, because their lives depend more upon these special powers than ours do. In contrast, human beings can out-wit them all because they have the resourcefulness of reason, combined with a beautiful mind, making them feel at home in many different fields. Animals, on the other hand, do not try to answer the ultimate questions of life because they cannot ask them. All they "know" is what their senses tell them.

People who deny that humans are superior to animals and plants, would say that all of them are equal—they exist—and that's all that matters. Whenever we evaluate some beings as "higher" than others, we read our own preferences into reality, but those preferences are not "real." Really? If you want to talk in terms of preferences, I would like to argue that to hold this view is a preference in itself, but that does not necessarily make it true or real. The opposite view is that there are real distinctions in the nature of things—for instance, humans differ from animals and plants. They all exist, but not on the same level. Denying this is another case of reductionism.

Just ask yourself the question why some apes are still swinging from trees, while humans are able to walk on the moon—something apes can't even think of doing. G. K. Chesterton said it well, "the more we really look at man as an animal, the less he will look like one." We need to face the fact that our inability to let animals be animals has something to do with our inability to let human beings be human beings.

4. At the level of organisms, we are degraded to *nothing but* a bundle of instincts as Sigmund Freud put it. That is how Freud ended up with a new form of reductionism: we are only a bundle of conflicting impulses, each one of which is striving for control, whereas we ourselves are just reduced to passive onlookers. If Freud were right, we would be fully at the mercy of our instincts. As a consequence, many kinds of behavior—especially morally good or wrong acts—have been turned into diseases, into some kind of pathology. Thus, grief became depression, apprehension became anxiety, rape became sexual frustration, religion became a child-

ish search for protection, and so on. What is in a word? Philosophy tells us: a world of differences!

No wonder, this form of reductionism would greatly affect our view on Man by turning us into a pack of meat, a meat-sack, or a meat-bag—a bundle of mere appetites and desires clamoring for satisfaction. Thinking along this line of thought, Marvin Minsky, a pioneer in the field of artificial intelligence, described human beings as mere "machines made of meat." It explains how Archie Bunker felt entitled to call his son-in-law "meat head." Again, reduction to this level is a legitimate technique, but when transformed into reductionism it becomes a devastating ideology. The consequences are clear: it turns human beings into cattle and cheap meat. It can eventually lead to concentration camps and unlimited experiments on human beings.

5. At the level of organs, we are degraded to *nothing but* a complex machinery. This is the level where human physiology operates. Reduction to this level is, again, a legitimate technique yielding great results. It is probably no coincidence, for instance, that many discoveries about the working of the human body were inspired by the latest technological contraptions of the time. The camera with its lens helped us understand the working of the human eye; bellows clarified how the lungs can do their work; pumps revealed what the heart does for blood circulation; computers threw some light on the working of the brain. In line with this view, food is to the body what fuel does to the steam-engine. No wonder

then, when things go wrong, we use technological devices to correct them: lenses, hearing aids, pacemakers, and prostheses—and rightly so.

We see here again what reduction can do for us, but when transformed into reductionism, it becomes a "nuts-and-bolts" ideology that reduces human beings to mere machine-like automata, which just follow all the physical laws of the Universe. The result of this outlook is quite predictable. Even in healthcare, we tend to think of a human being as a machine-like mechanism that needs to be "repaired" if it is not working properly, so we can fix or replace parts of it as if they were parts of a machine. However, looking at human beings *as if* they were machines does not *make* them machines. Needless to say that many aspects of human life have become obscured. To name just a couple: love has been replaced with machine-like interactions, and human freedom has been erased from this machinery. This makes us even think that the writings of William Shakespeare must have been secreted by the machinery of his brain.

6. At the level of the brain, we have been degraded to *nothing but* a pack of neurons. This is the level where neuroscience operates. Sure, reduction to this level is a legitimate technique, but when transformed into reductionism, it becomes a deceiving ideology. Although the human brain does indeed work with neurons, reductionism reduces the brain to a mere network of neurons, similar to the way a computer is a network of electrical components. The reasoning goes as

follows. Neurons either do or do not fire an electric impulse—in the same way as transistors either do or do not conduct an electric current. Seen this way, it looks as if the brain works like a computer works. It is the imagery of an automated machine, and nothing more.

The problem of this ideology, however, is that the real machines we compare human beings with are always the very product of those human beings. So, the question emerges as to who or what produced these "machines" that we call human beings? Did they make themselves? But that wouldn't make sense, for nothing can cause itself. As said earlier, to claim that a machine created itself is in fact philosophical nonsense, for such a machine would have to exist before it came into existence. The fact remains that looking at the brain *as if* it were a machine does not *make* it a machine.

7. At the level of cells, we have been degraded to *nothing but* a string of DNA. This is the level where geneticists feel at home. Of course, reduction to this level is a legitimate technique, but when transformed into reductionism, we end up with an ideology again. It has become a popular slogan among some scientists to make DNA the one and only "secret" of life. Someone like the American molecular biologist Walter Gilbert, for instance, had the audacity to claim that "when we have the complete sequence of the human genome we will know what it is to be human." Another one, the biol-

ogist E.O. Wilson, dared to exclaim, "People are DNA's way of making more DNA."

The problems with this ideology are manyfold. One, DNA in itself—for instance, in a test tube—cannot do anything. It is like a dead piece of text that needs to be "read and interpreted" to make sense. DNA needs at least the extended "machinery" of the cell to make its information available. Although DNA does hold vital information, the problem is that information cannot interpret itself. Problem two, whatever DNA does produce in the form of proteins is not complete yet but needs to be processed properly. That process does not seem to depend on DNA. Problem three, DNA cannot even copy itself on its own; it needs "help" to do so. Left alone, DNA degrades easily. Problem four, DNA may be able to create a replica of itself, but it cannot create itself from nothing. And there are several more objections against the slogan that DNA is the "secret" of life, belying the erroneous claim that life is *nothing but* DNA.

More in general, it can be said that DNA on its own cannot make itself work the way it does work. Making such a claim would lead us into a circular form of reasoning by assuming already what one attempts to prove. How can we step out of this pernicious circle? One way—perhaps the only reasonable one—would be to connect DNA and the information it carries to its Maker, God. The DNA expert Francis Collins, for instance, spoke in this context of DNA as "the language of God." It would be the language that preexists in the Mind of God before it can operate. That is where the "real secret" of DNA comes from.

8. At the level of molecules, we have been degraded to *nothing but* a mixture of basic chemical compounds. This is the level at which molecular biology, for instance, operates. Reduction to this level is certainly a legitimate technique, but not when transformed into reductionism. Ard A. Louis, University of Oxford professor of theoretical physics, makes fun of this form of reductionism by telling us our bodies have enough phosphor (P) for 2000 matches, enough iron (Fe) for one nail, enough chlorine (Cl) to disinfect a swimming pool, and enough fat to make 0.1 bars of soap. He suggests you tell this to a person whom you love, and then see what the response will be. The response will most likely be that love is more than a chemical reaction or a series of molecular interactions.

In short, we may be made of molecules, but molecules do not make us. We could actually reverse the roles: it is us who make molecules, not only in the industry, but even in our own bodies. For example, it is a normal part of life that cells in our bodies grow and then die. When 200 million of the 25 trillion red blood cells die each second, the 100 million hemoglobin molecules in each cell must be replaced immediately. With the order of a few thousand bytes of information in each hemoglobin, this is 10 thousand x 100 million x 200 million = 2×10^{20} bits of information per second—a million times more information processing than today's fastest computer CPU can accomplish. In short, our bodies make molecules as much as molecules make our bodies.

9. At the level of atoms, we have been degraded to *nothing but* predominantly empty space. This is the level where particle physicists feel at home. Again, reduction to this level is a legitimate technique, but it becomes a flawed ideology when transformed into reductionism. Think of this. We learn in school that an atom is composed of electrons and a nucleus. The electron is point-like and most of the mass of the atom is in the nucleus of the atom. In any given atom, the nucleus contains 99.9% of the mass of the entire atom. But its diameter is only 1/100,000th of the surrounding electron cloud. Given that the size of the atom is determined by the orbit of the outermost electron, the atom is thus 99.9999999999999% "open space" or "empty space" in the minds of many.

As the physicist Anthony Rizzi once put it, "If the nucleus were such that it was the size of a basketball, the 'edge' of the atom would be two miles away." It is thus erroneously believed, according to Rizzi, that the atom is filled with mostly "empty" or "open" space—which made a student say during an exam that atoms mostly consist of air! If that were really and literally true, though, then imagine the notion that objects like the human body are in fact collections of a huge number (10^{28}) of atoms. Rizzi again, "We would end up being mostly empty—or 'air' in the minds of some." Whether this is really "empty" space remains to be seen.

What all nine cases mentioned above have in common is that they are not scientific conclusions but merely an ideology of people who have fallen prey to nothing-but-ery. They confuse conclusions based on reduction with conclusions based on reductionism. The former set may be warranted on scientific grounds, but the later set never is. It is fair to say, for instance, that human beings are made of atoms, molecules, and cells, but that does not entitle us to claim that they are *nothing but* atoms, molecules, and cells. Reductionism is not something science shows us—it is rather something some scientists like to make us believe. I call that a "side effect" of science, or basically of what scientists have discovered. *All* beings are real—albeit on different levels of reality. A molecule, for instance, is as real as a human being. Reductionism has no right to claim that a molecule is more real than a human being.

S: Your separation of layers as found in the real world may look neat, but it makes me wonder whether these layers are really separated from each other. Or are they perhaps still connected to each other, and if so, how?

R: You are right on target with your question. On the one hand, there are many differences between the layers we talked about. Each layer has its own specific entities—we have, for instance, organisms, cells, and molecules, all of which are equally real. And these entities have their own properties—organisms can breathe, but cells and molecules do not. In other words, each level in the "hierarchy" may have properties not shown at other levels. This phenomenon is usually called *emergence*. The properties of water molecules, for instance, do not show up at the level of their components, the hydrogen and oxygen atoms. Something similar can be said about other layers; cells, for instance, can process oxygen, but DNA cannot. One could make the case that all sciences deal with hunks of "stuff," albeit of different sizes and different characteristics—stars, stones, cells, animals, human beings, molecules, atoms, particles, and so forth.

On the other hand, your question is also right on target as to how these layers are connected to each other. Or put in my terminology: how is emergence at a higher level possible? How can properties of a higher level "emerge" from properties at a lower level? Or to put this in terms of a whole with its parts: how can properties of the whole, at a higher level, emerge from properties of its parts, at a lower level. This calls for a brief philosophical discussion.

Sound philosophy has it that there appear to be at least three different kinds of "entities" in this world. First, we have *things*, sometimes called substances or entities, such as atoms, molecules, cells, organisms, societies, and the like. Next, we have *properties*, making certain things inorganic and others organic, or giving certain organisms plant-like characteristics and others animal-like characteristics. Finally, we have *relationships* between things, such as the relationship bees have to their colonies, or cells to an organism, or football players to their team.

I think these three distinctions are rather intuitive and obvious. However, since it is commonplace for philosophers to put question marks behind everything that regular people consider standard, it has also been a timeless temptation for some philosophers to deny the existence of one or two from the threesome mentioned above. Those, however, who accept all three categories as real, in line with Aristotle—and probably with most of us—could come up with the following characterization of a whole-part relationship: The *properties* of a whole (a *substance*) will typically depend upon *relations* among its parts (*substances*) as well as on *properties* of those individual parts.

Denying the existence of any of these three does not make them automatically disappear from reality. To distinguish all three of them, on the other hand, perfectly makes sense. Consider the simple example of two resistors in a circuit. Whether they are connected either in series or in parallel makes quite a difference for their total resistance. Hence, the properties of this system as a whole are not given merely by a complete knowledge of the properties of the two individual parts. Or take this chemical example: there are three different compounds—methoxyethane, propanol, and rubbing alcohol—with the same molecular formula, C_3H_8O, yet they have very different chemical properties. They are made up of identical atoms, but it is the arrangement of those atoms that determines which of the three compounds the substance happens to be.

Something similar can be said about living entities: it is a certain arrangement of certain molecules in a cell that makes life possible. Just putting a bunch of those molecules together in a test tube does not produce life. We showed this earlier with the example of DNA in cells. DNA may look like the "secret" of life, but the reverse is equally true: the life of a cell makes for the "secret" of DNA—that means, without the higher level of a cell, DNA is a "dead" or "lifeless" molecule and has lost its "secret." Take DNA out of its "context" and we have a useless molecule left.

What we may conclude from this is that science does have a dismantling approach. It is a powerful technique as long as we realize that when going to lower levels, we necessarily leave higher levels behind. Without that awareness we might unknowingly turn a successful reduction into a misleading reductionism. Scientists typically can and do dismantle reality as long as they know what they are doing. They are entitled to peel layers away, but they are not entitled to claim that those layers do not really exist.

When using reduction, we explain things by analyzing them into smaller and simpler parts. It is a technique that has undoubtedly yielded a rich harvest of discoveries about the universe. "Divide and conquer" is a basic scientific strategy for attacking any difficult problem: Take a large problem and divide it into smaller problems, then conquer the smaller problems one by one—for instance, problems in sociology and psychology might be analyzed in terms of biology, and biological problems might be analyzed in terms of biochemistry, and so on. As a means of analysis, then, reduction has certainly proven its value.

But this verdict changes completely when reduction leads to reductionism. If it were true that we are *nothing but* a statistic, star-dust, meat-bags, packs of neurons, strings of DNA, or bags of chemicals, then the search for a beautiful mind and soul would end before it could even get started. That's what makes nothing-but-ery such a devastating ideology.

The Cutting Knife of Science

S: You finally convinced me that dismantling is part of the way science operates with its strategy of "divide and conquer." But now you irritate me by claiming that science cuts reality apart with what you call a "cutting knife." You make it look like science mutilates its own objects of investigation.

R: No matter at what level we operate, science does use a "cutting knife" to cut things into pieces. Thus, they isolate, truncate, dismember, or dissect what is being studied. It is another way of using reduction: to explain things by cutting them down into its smaller and simpler parts. This technique looks very straightforward: take a large entity, use your cutting knife, and divide it into smaller parts, then conquer the smaller parts one by one. It is a technique that has undoubtedly yielded a rich harvest of discoveries about the natural and social world. As a means of analysis, then, the "cutting knife" has proven itself as a popular tool in the hands of scientists, for they feel they can understand the whole better when they know what the parts do.

But you are right, there is also a darker side to using this "cutting knife." It could be argued that complex systems cannot be split into their parts without mutilating them. To use a simple example, when you take an analog watch apart, you have no watch left, but only a bunch of pieces that do not do anything until you are able to put them back together in the proper way. In an analogous way, when you take a frog apart to see how it works, the first thing you have on your hands is a non-working, dead frog. Or put differently, when cutting apart a living frog to see what makes it tick, the "ticking" disappears, although all the pieces of the frog are still there. Obviously, we do not know yet how to put the pieces back together, in order to make the frog come to life again.

Aristotle famously wrote that a severed finger is no longer a finger. Apparently, there is something "hidden" in the original that the parts do not have. What then is this "something"? To put it in more general terms: how can something be more than the parts it is made of?

S: I know this idea was already recognized by Aristotle himself when he said, "The whole is more than the sum of its parts." But is that really true? Scientists have never been able to capture this mysterious "more" when it comes to living entities.

R: You are right, there has been quite some confusion on this issue. On the one hand, there seems to be something "hidden" in the whole that the parts do not have; on the other hand, we do not know what that "something" would be. Do we have reached a stalemate here? I do not think so. The answer to this problem can be found in what we discussed earlier when we mentioned the distinction between substances, properties, and relationships. Let us see what this terminology can do for us.

According to some, the *whole* (say, an organism like you or me) is nothing more than the sum total of its *parts* (say, its organs or cells). Many scientists would say that only the parts are presumably "real," while the rest is considered fiction. But that is basically ignoring the problem. Therefore, others would object to this and claim that a cell is certainly "more" than a mere collection of its molecules, or that a society is more than a collection of individuals—or more in general, that the whole is "more" than the "sum total" of its parts.

I would say these latter people make a good point. But the question is how we should take this "more" that is considered to be "hidden." If we take it as another substance with properties, we get into trouble. That happened, for instance, when some life scientists in the past introduced a mysterious substance, a so-called "life-force," to explain why living beings are alive—that is, more than the "sum total" of their non-living parts. They just postulated another substance.

What these people did not realize is that the parts of a whole also have a certain *relationship* to each other. Taking the whole apart destructs what we had before by breaking the relationships between the parts, hence the parts cannot contribute to the whole anymore. That is the reason why an isolated DNA molecule does not do anything unless it is part of a larger network of molecules located in the cell. This also explains, for instance, why viruses, which are DNA or RNA molecules packed inside a protein coat, cannot do anything until they penetrate, like a Trojan horse, the interior of a "living" cell where they utilize its machinery and thus become "part" of a larger "whole." That is also why a test-tube filled with all the molecules necessary for life does not produce life, because the proper relationships are missing.

So far, I used mostly biological examples, but something similar can be said about the situation in physics. Although I do not feel competent enough to discuss this issue much further, I think I can say the following. When dealing with atoms, it seems plausible to search for subatomic particles that make up the atom. Particle physicists have come up with atomic parts such as electrons, protons, and neutrons. In turn, they have found out that protons and neutrons can be further split into sub-particles made of quarks. Whether these are really "particles" in a literal sense is a different issue. Sometimes—that is, under certain experimental conditions—they do display particle-like behavior, at other times they display wave-like behavior. But the main point is that the relationships between the electrons, protons, and neutrons make the atoms react the way they do.

S: Doesn't this tell you that our universe is thoroughly material? Physics shows us that the ultimate constituents of reality are elementary, material particles and subparticles. What more evidence do we need to claim that everything is material and made up of elementary particles?

R: It depends on how you look at it. It is certainly true that we can study the material substrate of our world—that is something the reduction tool of science can do for us—but then one cannot just turn reduction as a legitimate technique into reductionism, which is an ideology. This ideology tells us that *everything* that exists is matter, or material, and that matter is *all* there is, so there is *nothing* else but matter. In this view, nothing can exist without the materials out of which it is made.

However, there is no logically safe way to reason from "there is matter in this world" to "there is *only* matter in this world" or to "*everything* in this world is matter." The problem to prove claims like these is that we are dealing here with so-called universal statements in logic. There are two types: universal positive statements ("*all* Xs are Ys") and universal negative statements ("*no* Xs are Ys"). Materialism claims both: *all* entities are material, as well as *no* entities are immaterial. Universal statements like these are hard, if not impossible, to prove (except in mathematics). The reason is that one would have to be all-knowing or receive divine revelation to be able to make such claims.

To use a simple example, it is indeed much easier to establish that there is a white swan somewhere on the earth than to prove that *all* swans are white. Alternatively, it is much easier to establish that there is a black swan somewhere on earth than to prove that *no* swans are black. To prove universal statements like these two, one would have to observe the earth in its entirety—from a viewpoint outside the earth. Well, the same problem holds for universal statements like "all entities are material" or "no entities are immaterial." They are universal statements about the entire universe. There is no way we could conclusively reason from "many" to "all." After having seen many material things, one cannot safely conclude that *all* things on earth are material—that conclusion is not logically justified. There is no way of knowing. One cannot even defend such arguments by saying they have worked so many times in the past, for that would be another example of moving from "many" to "all." This would take us on a never-ending search of proof.

Applied to reduction, it can be said that reduction can never prove that reductionism is a logical consequence of reduction itself, because it expands reduction to *all* possible cases. Therefore, no case of reduction can ever prove that reductionism must be true. One can certainly accept reduction, yet at the same time, reject reductionism. I have my reasons to do both. Therefore, I cannot accept the claim that matter is all there is.

The Matter of Matter

S: I am certainly ready to admit that the term "matter" is very ambiguous, especially so in physics. The term has in fact lost its prominent position in physics. The nuclear physicist Werner Heisenberg, for instance, put it this way, "[A]toms are not things." While the term "mass" is well-defined, the term "matter" is not. So, I must ask you: why would matter deserve extra attention in our discussion?

R: Scientists still keep referring to "matter" as the basic constituent of this world. You will find the term over and over again in their discussions and explanations. They speak in textbooks and articles of "the particles that constitute *matter* and radiation," or "the constituents of all *matter*," or "the nature of mass, energy, and *matter*," or "the fundamental subatomic constituents of *matter*," and so on. According to the subtitle of a certain textbook on particle physics, these elementary particles are considered to be at the "heart of the matter," so it is claimed. But also, scientists have come up with terms such as "dark matter" and "anti-matter" (which I will not further discuss here). In fact, the word "matter" is still all over the place in physics. It just carries on a long preexisting history.

As early as the 5th century BC, Democritus proposed that all matter was composed of small indivisible particles called atoms. Interestingly, the word atom is derived from the Greek word *atomos*, which means "uncuttable," so it was considered an ultimate entity that could no further be split or cut. Modern physics may have discovered that atoms can still be further split into subatomic particles, but the old idea of indivisible entities remains standing, although modified and adapted to new insights. Yet the prevailing idea is that we cannot cut things endlessly with the "cutting knife" of science. There are some elementary particles that cannot be split any further it seems. Thus, we may have reached what is considered the most basic "stuff" in the universe. How we define this "stuff" is still unclear: perhaps in terms of particles, or of waves, or of superstrings.

Let us provisionally define matter as anything that has mass and exists in space and time. Most laypeople have some intuitive idea of what "matter" stands for—they probably think of it as "stuff." That is exactly where its appeal lies. We are dealing here with the most noticeable elements in the world around us—indeed, "stuff." It can usually be seen, touched, heard, tasted, and smelled by our five senses or with tools that make them accessible to our senses. Studying this kind of "stuff" has been vital for our survival. We cannot disregard it; we cannot live without it; it is impossible to deny that almost all our technological achievements are based on it. Where would we be without "stuff" like this? That may indeed be true, yet this does not mean that matter is *all* there is.

S: I do not see why not. Didn't we learn at school that all things are composed of atoms, and that atoms are the building blocks of all matter. Nowadays I would say that subatomic particles are the ultimate building blocks of atoms, and therefore of matter. Since human beings are also made of atoms and subatomic particles, they must find their origin in matter as well.

R: I agree that we, too, are made of atoms and their subatomic particles—that is what reduction tells us—but I do not agree that we are nothing more than matter in the form of atoms and subatomic particles—for that amounts to the ideology of reductionism again. It is a particular form of reductionism that is usually called *materialism*.

The fact that the world we perceive through our senses and all the things we can picture are individual physical things or material embodiments gives indeed great credibility to the materialistic thesis that the world of real existences is entirely material, and that nothing immaterial really exists, because we do not have senses for it. Consequently, *all* things, including mental aspects and consciousness, must be the result of material interactions, thus making mental and conscious processes merely by-products of material processes—that is, by-products of the biochemistry of the human brain and nervous system, for example. The physical world is literally *all* there is, says this mantra. Materialism emphatically proclaims that nothing can exist without the materials out of which it is made. The physicist Carl Sagan worded this view with great flair when he said that the material world is "all that is, or ever was, or ever will be." Notice the word "all."

Obviously, materialism rejects religious faith and replaces it with its own "faith," faith in an ideology. However, it remains difficult for materialism to ignore non-material entities such as thoughts, beliefs, and intentions—materialism itself is in fact one of them. Are they less real than material entities? That is hard to believe. Keep in mind, for instance, how much Hitler's intentions, thoughts, beliefs, and his bigoted reasoning did shape human history. Were they less real than his gas chambers? Or picture what would happen if a majority of people were to decide to withdraw all their money from the bank. Are those decisions not real? In contrast, I would counter that reality is not a monolithic material entity. Yet, materialists dogmatically reduce or limit reality to material objects. This makes materialism a totalitarian ideology, for it allows no room for anything but itself. Can that be true?

Here is the kicker. If matter is indeed all there is, then one should wonder what the ideology of materialism itself is. Another piece of matter? Obviously, "matter" is a material entity, but it is hard to believe that "the matter of matter" is also a material entity. It is a *thought* issue, based on reasoning and arguments. Are thoughts also material? If they are, materialism has prevailed. But if they are not, then there must be more in this world than matter, thus undermining the core claim of materialism.

On the other hand, if thoughts are indeed material, then they, too, must be based on molecules, atoms, and subatomic particles. However, I do not see how thoughts can be material entities. Of course, we can record our thoughts with a Dictaphone, for instance, which makes them material. Or we can write our thoughts down on paper, which "materializes" them in another way. But what about the thoughts themselves? How can thoughts be material in and of themselves? How can the thought of prime numbers in mathematics, for instance, be material? How can the thought of non-existing entities such as centaurs be material?

Suddenly, we find ourselves in a non-material world where things are not large or small, light or heavy, hard or soft—but instead true or false, right or wrong, real or imaginary. Thoughts have no mass, no size, no color. In the world of molecules, nothing is true of false; one molecular structure cannot be truer or less true than another structure. In order to make any claims whatsoever, we need to validate our claims as being true, otherwise they are worth nothing. If we ourselves were nothing but molecules, this very statement that we are making here would not be worth more than its molecular origin, and neither would we ourselves. Such claims just defeat and destruct themselves. They cut off the very branch that the person who makes such claims was sitting on. I consider that suicide!

Another problem of materialism is its claim that "matter" is self-explanatory—as if it can explain its own existence. But how could matter make itself exist? That would be really magic. Nothing can make itself exist, for then it would have to exist before it came into existence, as we said earlier. Explaining matter in purely material terms is no explanation at all, but more like a circular argument in which one assumes what one attempts to prove. Material things cannot account for their own existence for the simple reason that matter is that which, by definition, already exists. Matter itself is merely something that must receive its existence from somewhere else. Matter cannot do so on its own.

That is the point where religious faith can, or even must, come in: God is the one who brings matter into existence and keeps it in existence. The late Harvard philosopher and mathematician Alfred North Whitehead once famously said that while math could count the atoms in an apple, it could not provide the apple itself.

S: I don't see the problem you mention! Materialists will just quickly respond that thoughts are in fact material as well. Thoughts are, so they say, just a certain pattern of electrical impulses in the neurons of our brains. That means materialism is still claiming victory.

R: I think you are undermining your own position, too. The curious thing is that you deny the existence of anything immaterial, but by doing so you also deny the existence of your very own denial. The truth of the matter is that materialism *assumes*, in advance, that nonmaterial objects cannot exist. But this assumption in fact turns out to be self-destructive, thus undermining its own nonmaterial assumption.

All I have said so far should make you realize that materialism is a form of reductionism that basically undermines itself. In other words, if I am certain that everything is material—physical, chemical, and so on—I would have no reason to suppose that my certainty is true—and hence I would have no reason to be certain that everything is material, and material only. Once I deny the existence of thoughts, I must also deny the very thought of materialism. But we are then still denying something, which in itself is an immaterial activity. We can think about sizes and colors of things, but the thoughts themselves do not have sizes and colors.

The amazing part is that we, as human beings, have the capacity to think and use thoughts. If man is only "little more than a chance deposit on the surface of the world, carelessly thrown up between two ice ages by the same forces that rust iron and ripen corn," as the historian Carl Becker once put it, then we should ask this scholar, "What is man that the electron should be mindful of him?"

The point I am trying to make is that human beings are more than their molecules, atoms, and subatomic particles. What this "more" looks like we will find out in the rest of this book. That will lead us to my claim that, in addition to our bodies, we also have a beautiful mind and soul. But before I do so, we need to address whether science has the qualifications to block religion from speaking about a beautiful mind and soul.

CHAPTER 3

When Science Becomes a Religion

Probably all scientists agree that science cannot deal with issues such as "mind" and "soul." Those terms do not exist in their vocabulary, because they are non-material entities—and science is supposed to deal with material entities that can be counted, measured, quantified, and dissected. Well, scientists are right, mind and soul do not qualify for a scientific approach.

But does that mean these issues are not real and do not even exist? That certainly does not follow. Yet, some scientists think they are entitled to go that far. They may even go much further than this: they believe science can actually block *religion* itself from speaking about non-material entities such as a beautiful mind and soul. Some of them might go as far as thinking they can replace religion with a new religion, the religion of science. Do they have a point?

Has Science Final Answers?

S: Religious people always tell me that religion has all the answers to our ultimate questions. Instead, I would tell them that it is science that has ultimate answers to all our questions. Maybe not yet, but I strongly believe all our

questions will eventually turn out to be scientific questions waiting for scientific confirmation and scientific answers.

R: What you say is a rather common opinion nowadays, also among many scientists. Here is one of them, the late astrophysicist Stephen Hawking, who once exclaimed, "[O]ur goal is a complete understanding of the events around us and of our own existence." Notice the words "complete" and "existence." Saying something like this exalts science to the status of a semi-religion that has final answers to *all* our questions.

Let me make a few comments on this position. My first comment would be this. The claim that "all questions have scientific answers" is an example of a *universal positive* ("all X's are Y's")—an unending proposition which is logically impossible to prove, as we discussed earlier. Even if there are many scientific answers to many of our questions, "many" does not equate to "all." It is at best a conviction, but not a safe, legitimate conclusion. This conviction would only be justified if we knew all the possible questions people might have, but also all the answers that science would come up with. It requires an all-knowing and all-present position—something religion assigns to God alone.

My second comment is that, as a matter of fact, science does not have answers to all the questions we might have.

This can be said about most questions that mathematicians, for example, have come up with. Finding the square root of a number cannot be done with a laboratory test. Finding the sum of the three angles of a triangle cannot be done in the lab either. And something similar can be said about many other questions. The most well-known question of this kind is "Does God exist?" As a matter of fact, God cannot be "trapped" by some ingenious kind of experiment, for God is not a material entity.

My third comment is that the reach of science is limited by the tools it uses. A telescope, for instance, can detect stars, whereas a microscope cannot; a microscope can reveal cells, but a telescope cannot. More in general, there are many questions for which science cannot provide answers, because science does not have the tools to get to them. Science can only deal with what can be quantified, measured, counted, and dissected, so it has no access to all that cannot be quantified, measured, counted, and dissected. This means that not all our questions have a scientific answer phrased in terms of particles, quantities, and equations. In short, science is "blind" for everything that is non-material.

My fourth comment is that science has its own window on the world, its own scientific point of view, its own perspective. However, there are many other windows, views, vistas, perspectives, or whatever you wish to call them. To claim that there is no other point of view than the "scientific" worldview, or that there is no corner of the universe, no dimension of reality, no feature of human existence beyond science's reach is not something science can prove on its own. In other words, it is not based on scientific evidence, but on a dogmatic, unshakable belief in the omnicompetence of science. It is a creed that makes for a semi-religion, because it can only be accepted on "faith."

My fifth comment is that many scientists, fascinated by their own science, also think they are free from any extra-scientific influences. Because they assume there is nothing more to material reality than what science tells us, they proclaim themselves "free" of any world-view, any viewpoints, any philosophy, or any values. They believe that there is in fact no worldview and no metaphysics in what they claim. If there is any "metaphysics" in their claims, they tell you, it is a meta-physics "reduced" to physics. All scientific truths are supposed to be "pure" and "naked" truths that only science has access to.

My sixth comment is that many answers science has come up with are not even final—they are only provisional until a better answer becomes available. In fact, no scientific statements are final and proven; instead, they always remain open to re-evaluation as new data is acquired and novel technologies emerge. Whenever there is new, strong, and convincing evidence in conflict with an existing theory, something needs to be adjusted. But what? Sometimes the evidence is put aside, sometimes other theories connected with the theory under investigation are revised, at other times parts of the theory are modified. What some like to call "proven" scientific knowledge is only proven until a new set of empirical data "disproves" what was previously considered "proven." So "empirical proof" is not cast in iron; whatever is true today may not be true tomorrow. Science is always a work in progress.

S: I still maintain that many parts of science can be considered "closed" and will not need revision anymore. I think of the theory of relativity, the Big Bang theory, the theory of continental drift, and so forth. Somehow science has achieved final results in those areas, I would say.

R: I am not so sure about that. We won't know whether these are final answers, because better answers might come up someday—but that is hard to predict ahead of time, but also hard to deny ahead of time. Perhaps the most telling story comes from Max Planck, the father of the quantum theory. He was told during his training that physics had been completed and finished. Well, it was he who, in 1900, laid the foundations for one of the greatest leaps in physics, the quantum revolution. In fact, what was arguably "finished" at the time is what we now call classical physics. Apparently, what was not finished was physics as such.

It is to be doubted whether it is in the nature of science ever to arrive at completion. Yet it remains a timeless temptation to claim that the unknown has been reduced to almost nothing. However, the magnitude of the unknown is, well… unknown! The late American physicist John A. Wheeler told his fellow scientists: "We live on an island surrounded by a sea of ignorance. As our island of knowledge grows, so does the shore of our ignorance."

Because of this open-endedness of science, we must also accept the possibility that certain "scientific facts" may turn out not to be facts at all. That is the reason why many current "scientific statements" may have to face the prospect of becoming outdated. It is a frequent saying that certain statements in widely used textbooks are not quite true, or no longer true at all; or that certain sections have become outdated, and so on. That is how science grows. A snapshot picture of science only shows us part of a much larger, continuing process at a certain point in time.

To think differently makes me think of one of Bill Keane's cartoons in which he has a little boy ask his Mom while doing his homework, "At some point, school will run out of things to teach me, right?" Unfortunately for the boy, it's a dream that won't come true. Unfortunately for scientists, the dream of final answers won't come true either. In general, science grows in fits and starts.

S: How can you say that some scientific facts may turn out not to be facts? Facts are facts, aren't they? You have consistently been defending the claim that facts are facts, but now you seem to have forgotten that.

R: The case is not that simple. As a human geneticist, I always think of the “scientific fact” that human beings have 48 chromosomes in their cells. That was considered a fact during the 1920s. Because the number was supposed to be 48, scientists did everything possible to make their counts 48. The number 48 remained a fact for over three decades until it turned out *not* to be a fact. The new discovery was 46 chromosomes, which is now the fact we find in current textbooks. Not amazingly, it always baffles non-scientists how you can make such a mistake in simple counting. How can there be disagreement if all one needs to do is counting properly? Well, the case is more complicated than that—as it usually is in science.

The problem is that chromosomes only form and appear around the time a cell divides into two new cells. To count them, we need a tissue that has enough dividing cells and then we must apply a hypotonic shock to make the chromosomes nicely spread so they can be counted. This technique was still not enough to make the new discovery of 46 chromosomes possible. Another stimulus came from the use of colchicine, which stops cells in their dividing stages, which increases the number of cells that can be used for counting.

Will this number of 46 chromosomes change again in the future? Most likely not, but there have been many more aspects of chromosomes that keep changing and tell us more and more details of what is inside the chromosomes, the DNA helix. New developments in DNA technology have been staggering. It is almost impossible to keep track of the latest developments. Is the end in sight? Of course not. As said before, the magnitude of the unknown is unknown! The co-discoverer of DNA, Francis Crick once said, "A theory that fits all the facts is bound to be wrong, as some of the facts will be wrong." He was right, but not precise enough in his statement. To express this idea more accurately, he should have said instead that facts cannot be wrong or false, but they may turn out *not* to be facts.

Has Science a Theory of Everything?

S: I learned in college that physics has a "theory of everything." I believe that claim to be true, because physics does pertain to everything. Hence, what physics tells us must be true of everything there is. Isn't that what a "theory of everything" proclaims?

R: Let me explain first why physics has received that reputation. Yes, organisms do contain cells, but not everything on earth does; yes, cells do contain DNA, but not everything does; yes, atoms do contain subatomic particles. However, and that is the salient point of your question, does this entitle us to claim that *everything* must contain subatomic particles? This last claim is quite popular among scientists. It seems to give physics, especially particle physics, a superior position among the sciences. It promotes the "omni-presence" of subatomic particles.

Is this position deserved, we should ask. On the one hand, it is, because it is based on reduction in *optima* forma. On the other hand, it is not, as it is also based on reductionism in *pessima* forma. Reductionism is a creed that needs to be toned down to a lesser statement: everything *material* contains subatomic particles. However, not everything is material, so we found out—just think of thoughts, for example. Thus, there might be things *not* made of subatomic particles. If everything were indeed material, physics would eventually be able to connect phenomena as disparate as quantum mechanical states and the songs of birds and the thoughts of humans and the culture of a society and the proofs of God's existence. This has led some to believe in the erroneous idea that everything in the universe can be explained based on equations and principles of physics. However, not everything is material, so we found out.

It is amazing to see how the creed of reductionism makes some scientists trumpet, for example, that physics always has the last word in observation, for the observers themselves are supposedly physical. But why not say then that psychology always has the last word, because these observers are interesting psychological objects as well? As a matter of fact, neither statement makes sense; observers are neither physical nor psychological, but they can indeed be studied from a physical, biological, psychological, or statistical viewpoint—

which is an entirely different issue. Even if physicists will ever be able to fulfill their dream of finding a "grand unified theory" (GUT), then this theory would definitely not be a theory of *all* there is in this world, but at best a theory of all *physical* phenomena, rather than a theory of *everything*.

Something similar can be said about the so-called "theory of everything" that you mentioned. If it is indeed about everything in this world, we should at least restrict "everything" to "everything that is material." So far, physicists have come up with a Grand Theory that unifies all physical theories—the so-called *Grand Unified Theory* (GUT). At this point in time, GUT is a physical "umbrella" theory that would unify the three non-gravitational forces—namely the electromagnetic force, the strong nuclear force, and the weak force—as aspects of one single underlying force. They are combined in the so-called Standard Model of particle physics.

There are still some questions left unanswered in this model. For example, combining quantum theory with Einsteinian gravity as well is more of a challenge. It is not yet reality, but only a dream that physicists have, despite many trials made on the "drawing board." Probably the best of these trials is *superstring* theory. We do not know whether superstring theory is the true theory, and we may never know, but many physicists believe they have good reasons to assume it is. A GUT that would also incorporate gravitational theory is sometimes referred to by physicists as a theory of everything (TOE)—which is an unfortunate choice of words and should not be taken literally as I explained already.

As to whether science can ever accomplish this goal of finding a "grand unified theory" in its widest interpretation, is an empirical question that only time can tell. But as I said earlier, even if or when science achieves this goal, its would-be "grand unified theory" (GUT) would definitely not qualify as a theory of *all* there is in this world—not even if it would include gravitational theory. It would at best be a theory of *physical* phenomena, not a theory of *everything*. There is no science of "all there is." A theory of everything (TOE) would also have to explain, for instance, why some people believe in it and some do not. And what would such a theory do with the thoughts of humans and the culture of a society and the proofs of God's existence? It is extremely unlikely, for instance, that such a scientific theory would have anything to say about the beautiful minds and souls human beings have. The truth of the matter is that one cannot give science the metaphysical power it does not possess—and that verdict includes any of its theories.

In short, it is important to realize that physics discloses only one of the many aspects of the material world, its physical aspect, thus providing physical answers to physical questions phrased in terms of physical causes and effects. Physics may be everywhere, but it surely is not *all* there is. Biology, for example, offers another perspective, providing biological answers to biological questions in terms of physical causes and biological functions. And psychology would study psychological questions in terms of psychological causes and effects. But there is no science of "all there is."

Limiting oneself to one particular perspective such as physics may be a valid strategic decision, which should not and cannot be changed into a metaphysical decision—reductionism. To quote Shakespeare, "There are more things in heaven and earth, Horatio, than are dreamt of in your philosophy." Having a beautiful mind and soul may be one of them.

S: No matter what you say, physics might still offer us a Grand Theory that could very well compete with the religious notion of creation and the existence of a creator. With this theory, we will not need a creator anymore—the Grand Theory takes care of everything.

R: Let me say it again: The Grand Theory describes how nature works at the level of subatomic particles. You seem to suggest that once we know how nature works, we do not need a Creator to make sure nature works the way it works—it does so automatically by following the "rules" of the Grand Theory and its Standard Model. But is such a claim really enough in itself?

What we are dealing with here is "the problem of *existence*," which is beyond the scope of physics. As the German philosopher, mathematician, and scientist G. W. Leibniz put it some three centuries ago: "Why is there something rather than nothing?" His point was that there must be an explanation (or reason) of why there is something rather than nothing. Without any further explanation, it is a mere riddle why there is something—that is, why something exists, rather than nothing. This question applies to anything that has come to exist, whether it is the universe itself or a Grand Theory, or anything else ranging from organisms to cells to subatomic particles.

Why is their existence, or existence in general, such a problem? The reason for this is quite basic and fundamental: nothing can make itself exist. Or put differently: everything needs a cause for coming into existence. Or perhaps even better and more accurate: everything that comes into existence must have a cause. This is not high-brow science, neither is it a form of sophisticated philosophy. Instead, we are dealing here with a principle that is universal and self-evident. We use and need it before science can even begin its search for causes and explanations of things that exist.

Let me say it again: nothing can just pop itself into existence; it *must* have a cause, because it does not and cannot have the power to make itself exist—"automagically," so to speak. Nothing can generate itself, just like children cannot generate themselves. How do we know this to be true? Well, it is really a self-evident truth: you cannot give your own existence to yourself or receive it from yourself—you cannot be your own father or mother. Nothing can bring itself into existence, or simply put, "nothing comes from nothing"—even plenty of nothing is still nothing. Not even the laws of nature could do the trick, as they presuppose the very existence of the Universe; therefore, they cannot be used to explain the existence of the Universe. Add to this that the First Law of Thermodynamics tells us that nothing in the universe (i.e., matter or energy) can pop into existence from nothing. As said earlier, that would even logically be impossible. Therefore, whatever has come into existence must have a cause outside itself that caused its existence.

Why is this still a riddle for so many? To say, as scientists like to do, that there is something because of something else—its cause, or explanation, that is—only shifts the question from one something to another something. Hence, we end up with a series of explanations—one explanation based on another explanation, with no end in sight. But we never reach a *sufficient* explanation or reason, not even by using a Grand Theory, for that poses the question as to how that very theory did come into existence.

Even if we explain a certain cause with a long series of other causes—a chain that may even be infinite—this long chain could not really explain the cause we want to find as an explanation. We would merely end up with an endless series of IOYs, but no real currency. It is the same with the entire chain of these causes: the chain is just floating in the air and keeps doing so until we find a "foundation" for it to rest on, or a "beam" for it to hang from. There is only one way to give any chain of causes a foundation to "rest" on or a beam to "hang" from. Leibniz spoke of a sufficient reason, which needs no further reason or explanation. It must be outside of this chain of explanations and must be found in a necessary Being bearing the reason for its existence within itself.

Only a necessary, self-explanatory Being can explain with "sufficient reason" why this world exists. This is the *First Cause* St. Thomas Aquinas spoke of a few centuries earlier—not a cause working inside the chain but a Cause operating from outside of the chain. This First Cause is the Being that religious people call God—and no Grand Theory can replace this God. Without God, nothing would have a firm foundation and could not even exist.

S: The Grand Theory is a theory about *matter*. Why would matter itself not qualify as the ultimate cause and explanation of everything that exists in this universe?

R: There goes your ideology of materialism again. Even matter cannot be the cause of its own existence. The idea that matter can cause itself and explain its own existence has rightly been caricatured by the Boston College philosopher Peter Kreeft as a magical "pop theory" that has things pop into existence without any cause. That violates again the self-evident truth that nothing can generate itself, or produce itself, for it would have to exist before it came into existence—which is logically impossible. Even a Higgs boson, for instance, cannot just pop itself into existence; it must have a cause, because it does not and cannot have the power to make itself exist. To think differently would even violate the First Law of Thermodynamics.

We would still need "a necessary, self-explanatory Being," God. This Being cannot be replaced by the Grand Theory, for that theory cannot explain its own existence, nor can it explain why the universe exists. The laws of the Grand Theory can never explain the existence of the universe, for before the universe existed, we would have to posit these laws of physics—which are ultimately the set of laws which govern the existing universe. People who deny this are actually saying that laws which have meaning only in the context of an existing universe can generate that universe, including the laws of nature, all by themselves before either exists. This would be a form of circular reasoning, assuming already what one attempts to prove. It would also make for an absurd conclusion.

What we do find in this universe is a specific "law and order." Yet, the universe does not have to behave the way it does. Only a self-explanatory, necessary Being, God, can explain why the laws of nature are the way they are. However, if there is no inherent necessity for the universe to exist, then the universe, including everything in it, is not self-explanatory and therefore must find an explanation outside of itself. Only God is the explanation of the "law and order" in this universe. Without God nothing could even exist; there would in fact be no matter, no Grand Theory, not even any science. That is the reason why science can never replace religion. As Chesterton put it, "God is like the sun; you cannot look at it, but without it you cannot look at anything else."

The Priests of Science

S: If I go along with you, for the sake of argument, and admit that God created everything, then he must have created scientists, too. That should give them a special place in God's creation and that should give science an aura of divine sanction. Don't you think so?

R: I do not think scientists deserve a status different from other people. They were not given a superior position in God's creation. If they think otherwise, they have given themselves a special kind of superiority. The late Dutch physicist Hendrik Casimir—the Casimir effect of quantum-mechanical attraction was named after him—once said, "We have made science our God." Instead of placing science under God, they made science their god. Indeed, science has become a semi-religion of which the scientists are considered the "priests." It is not so much science that is to blame for this, but the ideologies of reductionism and materialism that are still quite rampant among many scientists.

I admit, science has one of the most impressive track records in human history. It is a success story that keeps persistently adding new achievements to the list, with no end in sight. Hence, it shouldn't come as a surprise that science has given us reason, not only for high hopes, but also for extravagant claims. Science and scientific knowledge have thus been endowed with the status of a new, comprehensive philosophy of life, close to a new religion. Some scientists even have the audacity to use science as a weapon to attack whoever disagrees with them. They say, "If you attack me, you attack science." One of the latest examples was Anthony Fauci, director of the U.S. National Institute of Allergy and Infectious Diseases (NIAID). Can that really be a valid defense? Some scientists tend to forget that many scientific claims had to be abandoned, or at least modified, in the past and will be in the future. There is nothing sacred about science—it is a methodology, not a religion.

Nevertheless, in the minds of many nowadays, the "building" that houses science has become a "temple" of science. Many still see this image in the way the Massachusetts Institute of Technology (MIT) was built in the early 1900s—with its impressive pillars. Although it looks like the Greek temple of a new "religion," it must be said it was not designed for that reason. If you want a closer resemblance of "a temple of religion," try Cambridge in England, where guys like Dawkins and Hawking are given a prominent platform, treating its scientists as little "gods." They love to trumpet their theory about a conflict between science and religion, which we talked about earlier.

Scientists with this kind of outlook think that they have scientific answers to all questions, including religious questions. Unwavering belief in the omni-competence of science is a modern form of superstition. This belief is something they cannot really prove of course, but it is a conviction they hold on to for dear life. Some of them have tried extremely hard, for instance, to replace the religious issue of creation with a "scientific" alternative. I think the most telling case of this travesty is the scientific replacement of the religious belief that the universe was created "out of nothing."

The religious belief of "creation out of nothing" dates back to at least 1215 when the Fourth Lateran Council officially taught that, at the beginning of time, the universe was "created out of nothing." What the Council meant by this statement is that God made the universe without making it out of anything else—and certainly not out of pre-existing matter, as some religions and Platonic philosophers have it. Creation has everything to do with the philosophical and theological question as to what makes things exist at all. Without God's creation, the universe could not even exist. Existence is received, comes from God, and requires God's creation and preservation. That is the centuries-old Judeo-Christian belief.

How then can some scientists—such as Stephen Hawking, Carl Sagan, and Alexander Vilenkin, to name just a few—claim that they have a scientific version of "creation out of nothing"? Well, they refer to what is called "quantum tunneling from nothing," which makes subatomic particles emerge and disappear spontaneously in vacuums in laboratories. They seem to have the facts on their side. When an "empty" space is filled with an electric field, but without particles, there is a chance that suddenly an electron-positron pair will pop out of empty space. On the other hand, the reverse can occur too: when an electron and a positron collide, they can "an-nihil-ate" and thus change into "nothing" [*nihil*]. They describe this "void" or "nothing" in mathematical terms with a wave function—which they consider the quantum gravity equivalent of the quantum vacuum in quantum field theory. From here they postulate that the Big Bang could also be the result of a random quantum fluctuation in a vacuum—a spaceless and timeless void. In translation: the universe came "out of nothing."

S: I must say their case looks quite strong to me. It is based on experimental, scientific concepts. What more could one wish for? If their case is true, then there is indeed nothing left for a creator to do. This may not be a case where science and religion are in serious conflict with each other, but they certainly are in competition with each other. In this case, science seems to be the winner.

R: I am not so sure that science can claim its own version of "creation out of nothing." The main problem is that "nothing" in religion is very different from "nothing" in science. They both use the same word for two quite different concepts. When philosophy and religion speak of "nothing" during creation, they mean literally "nothing"—a no-thing. It is not a highly unusual kind of exotic "stuff" that is more difficult to observe or measure than other things; it is not some kind of element that has not found a position yet in science's Periodic System; it is in no way a material thing that can change into something else, but it is in fact the *absence* of anything.

Therefore, "creation out of nothing" does not mean changing a no-thing into a some-thing, or changing something into something else—like chemists change water into hydrogen and oxygen. Instead, it means bringing things into being before they can even exist and change. We should not forget that the beginning of the universe coincides with the beginning of space and time. In other words, there is no way we can talk about what happened *before* the beginning of the universe, before the Big Bang. There is no "before" here, and therefore nothing for science to study. That's why the astrophysicist Robert Jastrow could say, "At this moment it seems as though science will never be able to raise the curtain on the mystery of creation." This calls for a divine explanation, not a scientific one.

In science, on the other hand, the situation is entirely different. When scientists speak of "nothing" in their supposedly scientific version of "creation out of nothing," they are not really speaking of "nothing" in the literal sense as used in philosophy and religion. When an electron and a positron collide in the lab, they can indeed "an-nihil-ate" and thus change into "nothing," but this "nothing" is not really nothing in the literal sense. What really happens is this: when these particles annihilate, they emit a burst of energetic photons—which is certainly not "nothing" in a philosophical or religious sense.

We must also face the fact that when scientists call a vacuum "empty," they are again using a deceiving kind of terminology. Even with the most modern techniques, we have never been able to pump everything away from a so-called vacuum. Scientists are getting better at it, but never has a vacuum become literally "empty." Since matter is basically a form of energy, there is always some matter or energy left in a vacuum. A vacuum, or so-called "empty" space, has energy and tension, it can bend and warp, so it is unquestionably something. This means that quantum particle generation requires pre-existing energy—again, a far cry from nothing. Besides, we need to stress again that energy could not have popped into existence without violating the First Law of Thermodynamics.

To put it differently, "quantum tunneling" causes a "system" to change from one "state" (an electric field without particles) into another "state" (a changed electric field with two particles). These two states are different "states," but of the same "system." However, it must be realized that a pair of particles does not suddenly appear out of literally "nothing," but out of an electric field of an existing "system"—which, again, is not "nothing" in a philosophical or religious sense. Apparently, in quantum tunneling, we are dealing with an existing "system" that has a set of possible "states." Obviously, a state is a state—a specific "state" of a specific, complicated quantum "system" governed by definite quantum laws.

To explain the idea of different states, Stephen Barr uses the striking analogy of having a bank account. Even when we have "nothing" in the bank, we still have a bank account, with all that comes with it, but it happens to be in a "no-money state" for us. Needless to say that this kind of "nothing" is different from having no bank account at all.

S: You told me before that science is never final. If you are right, then I could make the case that the laws of quantum physics may have to be revised in time so they can explain "creation out of nothing."

R: Even if your objection is valid, then there is still no way to escape from the ultimate key-question: why is there something rather than nothing? If it were indeed true that the universe owes its origin to quantum theory, then quantum theory must have existed before the universe came into existence. Hence, the next question is bound to be: where did the laws of quantum theory come from? To claim that they came "out of nothing" is a moot answer—not an explanation but a belief. These laws are not self-explanatory and cannot just "automagically" pop themselves into existence, as little as matter and energy can.

The truth of the matter is that "creation out of nothing" has everything to do with the philosophical and theological question as to why things exist at all—that is, before they can even undergo any change. Therefore, creation—but not the Big Bang—is the reason why there is something rather than nothing, including something such as the laws of physics and quantum physics. Only God can create something from literally nothing. Only God can make things come into existence and call them from nothing into existence.

The bottom line of this discussion is that some scientists may want to replace the one God of religion with the many gods of all-knowing scientists. By doing so, they create a semi-religion, but the truth of the matter is that science is not a religion and cannot replace religion. There is much more to life than science can offer us. All the above reflections, insights, and conclusions are only possible because each one of us has a beautiful mind and soul.

CHAPTER 4

Can Science Remove the Soul from the Body?

At the beginning of this book, we discussed that there is no conflict between science and religion. But now we seem to be facing one. Religion speaks of the twosome "body and soul," but science would reject any mention about what religion calls the "soul." Do we really have a conflict here? If so, science is most likely nowadays to come out as the winner based on a majority vote.

S: Mentioning the term "soul" is like placing a timebomb into the discussion. The idea of a "soul" puts a ghost into the machine of our body. That is dangerous, even scary. Science should immediately take it out again.

R: I must say the notion of a soul has repeatedly been contaminated in the history of philosophy. It was the philosopher, mathematician, and scientist René Descartes who compared the human mind and soul with pilots in their ships, suggesting that they pilot everything the body does. In this Cartesian view, a pilot can be without a ship, and a ship can be without a pilot. But it also implies that "pilots" are in charge of their "ships" and can use them in whatever way they want. This view has become known as mind-body dualism and has received a bad press.

As a matter of fact, the Cartesian view has always had quite an appeal to many people. A popular theory in vogue for a long time was that the human mind works with something like an "inner observer" or "mental agent"—something like a "man in the machine." Consider watching a movie: viewers see the images as something separate from themselves projected on the screen. Could we perhaps explain this by supposing that the light from the screen forms images on the retina which are then transferred to the visual cortex where they are scanned next by an "inner observer" located in the brain behind the cortex? Or when people give commands to their body, could it possibly be that some "mental agent" is steering the body like a piano player plays the piano?

No, not really. If this theory were correct, we would end up with "an observer in an observer" or "a piano player inside a piano player"—something similar to a set of Russian nesting dolls. You feel it coming—infinite regress—a series of never-ending statements, where the validity of one statement depends on the validity of the one which follows and/or precedes it, and so on ad infinitum. They sometimes call this the *homunculus* (little man) theory—explaining a phenomenon in terms of the very phenomenon that it is supposed to explain. Explaining "man" by assuming a "man in the man-machine" is a similar thing—it leads us nowhere. This made many conclude we cannot take the mind as an agent hidden somewhere in the body, or as "a man in the man-machine." From this they conclude the brain does not harbor any detectable agent who plays the "piano" of the body or steers the "ship" of the body.

Yet, scientists—and neuroscientists in particular—would like to find a part of the brain that controls the brain itself. It is a well-known fact that parts of the brain do "mirror," or represent, certain parts of the body. Since scientists always search for material causes, they have also been searching for a part of the brain that regulates the brain itself. But no such physical brain part in charge of the brain has ever been found; the brain does not seem to have a part that mirrors itself. If that is true, then the mind is eluding the knife of the dissecting neuroscientist—or, as some of them say, the mind just does not exist. To put it in a catch-phrase, brain-surgeons are not mind-surgeons. The brain is all there is to it, so there is no longer any "room" left for the mind, let alone the soul.

Therefore, the dualism of body and soul was soon replaced with the monism of the brain. But as Max Bennett and Peter Hacker have convincingly argued, the homunculus fallacy keeps coming back, even then, albeit in another form. Now the homunculus is no longer a soul, but a brain, which is supposed to "process information," to "map the world," to "construct a picture" of reality, and so on. Oddly enough, this brings the *homunculus* back in a hidden way, for all these latter expressions can only be understood because they describe conscious processes with which we are already familiar. To describe the resulting form of "brain-science" as an explanation of consciousness is therefore questionable, as it merely reads back into the explanation the feature that needs to be explained. It creates the false impression that self-awareness is a feature of the brain. It may still be considered "mind-reading," but now disguised as "brain-science."

S: I want to mention that the Cartesian view has also been under heavy attack from a philosopher such as Gilbert Ryle who described it as a "ghost in the machine" kind of philosophy in which all mental activity is supposed to run parallel to physical action. Doesn't Ryle have a valid point here?

R: True, Ryle did dismantle this experience as something spurious. He actually used the term "category mistake," in which things belonging to a particular category are presented as if they belong to another category. "Mind" and "body" are not terms of the same logical category, he says. One of the examples he mentions to show this is the following. Someone asks at the end of a tour through all the facilities and laboratories of Oxford University, "Now, where is the University?" That is certainly an odd question. The error this person makes is presuming that the University is part of the category "buildings." Obviously, the University itself is not another building.

According to Ryle, something similar holds for concepts such as "I" and "self"—and, I might add, for the concepts of "mind" and "soul." You cannot ask, after having seen all the parts of a body, "Now, where is I?" Therefore, according to Ryle, separating "I" and "self" from the body is a category mistake, which leads to some kind of "a ghost in the machine" theory. Although this discussion relates first of all to the human mind, it can as well be applied to the human soul.

At first sight, this objection seems to remove the concept of "soul," and of related terms, from scientific discourse, or from any kind of discourse, as being a category mistake—once and forever. But on a closer look, this may not be quite true. Ryle may think there is no university over and beyond the buildings of Oxford University. However, if there were no university in Oxford, those buildings would not be connected the way they are, or they might not even exist. Of course, the university is not another building; that would indeed be a "category mistake." Yet, there is something very real about the university itself that unites all the buildings we can see and visit. And the same could be said about the mind and soul. If we try to get rid of them, we lose something that "unites" what is going on in the body. What we "see" can be best explained by what we do not "see," yet is still there. Ryle cannot take this insight away from us, unless we join materialists who only acknowledge buildings as something "real."

When we ask questions about our own existence—such as "Who am I?" or "Why am I here?"—no scientific answer could possibly do the job. From the fact that two specific parents had a child, for example, it does not logically follow that this child is or became a new "I" or a new "self." We may very well need the soul, but even more so God, to explain why each one of us exists, and what it is that makes us to be who we are. We cannot put our finger on what this "I" or "me" or "self" stands for. It is like my shadow—always a pace ahead of me, leaving open what the next step will be. The shadow of oneself evades capture, and yet is never very far ahead.

Ryle' s line of reasoning that rejects the existence of a mind or soul is only valid if one adheres to the philosophy of materialism, which we questioned earlier in this book. If one only accepts the existence of material entities, then anything immaterial such as the soul—or a university, for that matter—must be a "ghost." Seen that way, the soul ends up as a "ghost" in the material entity of a machine. But then I ask you: what gives us the right to deny the existence of non-material entities? Not science, of course. Scientists may adhere to materialism, but such a belief is an assumption they start from, which is not the outcome of any scientific investigations. It's an ideology, not a scientific conclusion.

True, the soul is not something that can be measured or weighed. Weighing a soul is indeed hard to do, if not impossible. Failure to distinguish the immaterial soul from the material body made Duncan MacDougall, an early 20th-century physician in Haverhill, MA, measure how much mass a human body had lost after the soul had departed the body upon death. He came up with a weight of 21 grams! Although he never mentioned if some souls might be lighter or heavier than others, he did assume that souls do have weight. Where he went wrong is in mistaking the soul for something material, which led him to an absurd conclusion.

There is no way science can prove or disprove the existence of the soul. When George Bernard Shaw triumphantly exclaimed, “We know now that the soul is the body, and the body the soul,” one should have asked him how he knew that. Certainly not from science. Science has indeed great tools and techniques to detect material things, but it is inadequate to detect immaterial things. We cannot conclude from this that immaterial things do not exist. An instrument can only detect what it is designed to detect—the rest is beyond its reach. Because science detects only material things, it is “blind” for anything non-material. We cannot claim, for instance, that the square root of sixteen does not exist because we cannot find it under the microscope.

Therefore, anything immaterial—which includes the soul as well—is necessarily on science’s “blind spot” and thus is off-limits for science. When scientists still proclaim that religious people have a deep-seated need to believe in a soul, we could as well proclaim that those scientists have a deep-seated need *not* to believe in a soul. If you still maintain that the soul does not exist because it is immaterial, then think again: this very belief is worth nothing because it is in itself immaterial too. But I think I stressed that point enough.

I would conclude from this that we are still entitled to say that the soul is not identical to the body. We should affirm now, contrary to what George Bernard Shaw said, that the soul is *not* the body, and that the body is *not* the soul. In short, we are not just "automated machines" or "machines made of meat." There must be much more to the "machine" we are supposed to be. We may look at the body as a machine but that in itself does not *make* it a machine. To claim otherwise amounts to reductionism.

Where Does the Soul Idea Come from?

S: I still don't like your calling in a concept that is so tightly associated with religion. Religion loves to mention the soul, but that does not make the soul a reality.

R: True, when people hear the word "soul," they often think it comes from religion. In a way it does, but it can also be traced back to ancient Greek philosophy, to philosophers such as Socrates, Plato, and Aristotle, and soon after, it would place an important role in Christian religion. According to St. Thomas Aquinas, in line with Aristotle, all living beings have a soul (*anima* in Latin). One might say "soul" is how Aquinas explains life. When someone wants to know why an apple falls to the ground, we mention gravity nowadays; gravity explains why things fall. In the same way, when someone asks why an organism is alive, whereas a rock is not, we could mention the soul as an explanation for the difference between the two. "Soul" explains why things are alive. Therefore, when a plant or animal dies, its soul is gone too.

More specifically, Aquinas speaks of a "vegetative" soul for plants, a "sensitive" soul for non-human animals, and a "rational" soul for humans. This may sound like antiquated philosophy, overhauled by modern science, but that impression may be deceiving. It is actually an example of common-sense philosophy. Plants have a vegetative soul, which comes with growth, metabolism, and procreation; animals have a sensitive soul, with locomotion and perception; and human beings have a rational soul, with reason and intellect. These are "nested" in the sense that anything that has a higher degree of soul also has all the lower degrees. All living things grow, nourish themselves, and reproduce. Animals not only have those activities, but they also move and perceive. Humans also use reason, but they do all the above as well. Aquinas's ancient views still nicely tie in with current scientific insights.

All of this is based on the Aquinian (and Aristotelian) idea that every being is a unity of *matter* and *form*. Each and every individual thing is composed not only of matter [*materia*] but also of form [*forma*]. These two cannot exist without each other, but do not take the word "matter" in the way materialism does.

S: But if all life is outfitted with soul, what then makes a human soul so different? What sets the human soul apart from the other souls?

R: As a matter of fact, there is an important difference between the three types of souls. Vegetative and sensitive souls depend entirely on matter for their operation—for instance, photosynthesis requires leaves, and digestion is done in intestines. However, whereas plant souls and animal souls depend on matter, the human soul must exceed matter because it is able to grasp and process abstract ideas such as "circle," "truth," or "justice" which do not have length, width, or height. This requires the human faculty of intellect, which makes the human soul vastly different from plant and animal souls.

The human intellect is essentially *immaterial*, not requiring any bodily organ for its operation. We do need sensory organs for perception and sensation—to see a particular circle, for instance—but thinking about circles in general goes far beyond that one particular circle we are looking at. Perceived objects are never perfect circles, whereas the thought of a circle is about a perfect circle, without being a circle itself. Any material circle is always and only an approximation of perfect circularity. Therefore, the thought of circularity cannot be material itself. This makes the human intellect essentially *immaterial*, not requiring any bodily organ for its operation (other than the senses which provide the material input). Thoughts are universal and immaterial, whereas perceived objects are particular and material.

It is the soul of human beings that sets them apart from the rest of the animal world. The reason is twofold: the human soul comes with an intellectual part, the *mind*, as well as a voluntary part, the *will*. The intellectual part enables human beings to have thoughts and to think properly. The voluntary part enables them to have intentions and to act freely. *Rationality* comes with the *intellectual* part and makes each one of us an "animal rationale," so that our thoughts can be judged as true or false. *Morality*, on the other hand, comes with the *voluntary* part and makes us beings who live by a moral code, so that our actions can be judged as good or bad. Without a soul, human beings would no longer be rational beings and moral beings; they would no longer be able to have thoughts and intentions. To put it bluntly, they would be mere "brutes" like the rest of the animal world.

S: I see dualism creeping in again, but now as body-and-soul dualism, remarkably similar to the mind-and-body dualism of Descartes with all its problems.

R: The Christian view of body-and-soul in fact rejects the dualism of Descartes. Inspired by Aristotle and St. Thomas Aquinas, the soul is considered to be the substantial *form* of the body. In the Aquinian view, form and matter make up a substantial unity; one cannot have form without matter, nor matter without some form.

Applied to the relationship between body and soul, this would indicate that the "form" of the soul gives a specific existence to the "matter" of the body. Body and soul, or spirit and flesh, are two sides of the same coin; the soul is the form of the body, or put differently, the spirit is the form of the flesh. In saying this, Aquinas embraces the total reality of the human person as a living composite of spirit and matter, without emphasizing one element to the detriment of the other. In other words, we can *tell* body and soul apart, but we cannot *set* them apart. Therefore, we certainly do not have Cartesian dualism here.

The Catholic Church confirms this view in her Catechism (#365): "The unity of soul and body is so profound that one has to consider the soul to be the 'form' of the body: i.e., it is because of its spiritual soul that the body made of matter becomes a living, human body." Since a soul is "fitted" to its body similar to the way a key is fitted to its lock, a human soul can only "in-form" a body that has a brain complex enough to deal with and process human faculties such as language, rationality, morality, self-awareness, and religion. These faculties do not seem to come from the body but rather from the soul—in other words, they do not come through the umbilical cord. They are essentially immaterial, not requiring any bodily organ for their operation.

In this view, there is no human body without a human soul (unless it is a corpse), and there is no soul without a body (except temporarily, until the resurrection after death). A person is a unity of a human body and a human soul. This makes our bodies an essential part of who we are—which is basically a common-sense idea, and certainly not Cartesian dualism.

How Can Body and Soul Work Together?

S: You call the unity of body and soul a "common-sense idea"? In contrast, I would say that it makes more common-sense to say that the soul lives *inside* the body—which is the very dualism again that you have been arguing against.

R: If the soul would really live inside the body, it is hard to see how the two could communicate, for there is no "interface" where they could connect with each other. Descartes always struggled with this problem himself. Obviously, this kind of dualism calls for a "medium," some sort of transmission process between mind and body. Locating this "medium" or "interface" in the pineal gland, as Descartes suggested, does not solve the problem, because an immaterial soul cannot possibly contact any material part of the body, not even a pineal gland. The problem remains, though: how can the soul move or control the material entity of the body if it is in no way material itself?

In response to that question, I would argue that body and soul make for a tight unity—we can, as said earlier, tell them apart but not set them apart. The fact that we can distinguish a flame's heat from its light does not mean that we can separate the heat from the light. Neither can we separate the soul from the body. The soul feels what the body goes through, and the body shows what the soul goes through. When the body hurts, the soul hurts, and when the soul hurts, the body hurts. The unity between body and soul is so strong that we can even speak of "psychosomatic" symptoms that come from this twosome combined. Due to this tight relationship, we grieve with the soul, and the body shows how we grieve.

The fact that we don't quite understand how body and soul can work together does not mean we are stuck in an unacceptable situation. Something similar can even be found in physics. Not too long ago, quite unexpectedly, physicists had to face the fact that at the quantum level it is impossible to predict accurately where a particular particle will be found until a human "observer" makes a physical observation. In other words, physics could no longer be formulated without reference to human "observers." Suddenly, physics had been invaded by people "who make observations."

These people do not make for another physical system but introduce the "knowledge" of observers and their minds into the physical system. They cannot be absorbed into the physical systems they study, for quantum physics requires the knowledge of observers outside the system they study. That is as shocking as introducing a soul into the body.

Because souls are immaterial entities, we did not inherit our souls from our parents, let alone from some non-human ancestors, nor did our souls come with our DNA, nor did they come through the umbilical cord. Science knows nothing about souls and, of course, religion on its own knows basically nothing about bodies. But together they tell us a more comprehensive story. In other words, the soul is not part of the biological domain and vocabulary—or to put it briefly, the soul is not the body. Yet, the soul is necessary for us to be human.

Let me put it in this way: if we are looking for a key to understanding ourselves as human beings, then it will not only be in terms of matter and body, but it must also be in terms of soul and spirit. Since the soul does not occur on the physical, scientific map of the body, it may seem to be nowhere in the chain of bodily activities, yet it is the "soul" of it all and pervades the entire body; it is "part" of everything the body does without being a bodily part. The soul is "all-present" in the body, without being specifically located anywhere in the body.

We must thank St. Thomas Aquinas for stressing this tight unity. He knows and accepts Aristotle's assertion in *De Anima* that it is as pointless to ask whether soul and body are one as it is to ask whether the seal and the wax are one. They unquestionably are in the sense that the seal is the shape that is imprinted in the wax and makes it the kind of waxen thing it is. But obviously, they are also not the same thing, as the seal is not identical to the wax; they are as different as the body and the soul are—yet, we can and do distinguish them.

We have here a philosophy of monism in the sense that the material and immaterial are tightly united so that one cannot be without the other. Aquinas keeps repeating, "It's not my eye that sees, but it is *I* who sees; it is not my intellect that thinks, but it is *I* who thinks." This "I" comes from the unity of body and soul. Consequently, the twosome of body and soul remains standing and needs to be kept united for us to remain standing. Science tends to isolate the body from the person, "I," but the body is always the body of a *person*. And the same should be said about the soul—it is always a person's soul. It is not only the body but also the soul that determines a person's identity.

S: You mention as person's identity. How can the soul ever determine my identity? I can see that my body can do so, but I do not see how my soul could. You are asking too much from me as a scientist.

R: The soul is often identified with "me" or with my "self." In a way that makes sense. When doing so, it is important to notice that the "self" or "me" of a person remains the same, even though all the cells of one's body keep changing on a perpetual basis. The next minute, my body is no longer the way it was a minute ago. That makes it hard to see how a person's body can determine a person's identity. Because the elements of a person's body are being continually changed by eating and drinking, by wear and tear, they themselves cannot be the source of that person's "self."

Even losing or replacing some parts of my body may happen without losing my "self." I can get a new heart, liver, lungs; I can also get knee, hip, and ankle replacements; I can receive prostheses for hands and feet, arms and legs, and so on. Yet, I know I am still my own old "self," although perhaps suffering a bit more from those replacements. Even when cancer spreads in my body, I am still the same person with the same soul and the same "me." In that sense, the soul is a "fact" in life, comparable to the fact that the earth is not flat. The soul either exists or it does not—if it does, then it is a factual reality. We all know we do have a body—which is a fact hardly anyone disputes—but we should also know we have a soul—that is a fact, too, no matter whether we like it or not, or whether we believe it or not, or whether science accepts it or not.

It is the soul, not the body, that asks, "Why me?"—for instance, when something awful happens to me. The body has no "me," for it is just a collection of ever changing cells, molecules, and atoms. The cells, molecules, and atoms we were born with are practically all gone when we grow older. The only "thing" that stays the same is my own self, my soul. That's the very "me" which is asking the question, "Why me?" Just the fact that I can ask this question tells me already I am more than a body. That's undeniably just a matter of fact, too, no matter what someone like Ryle tells us!

When asking myself the question "Who am I?" I certainly do not do so from a desire to find out my own name, gender, age, etc. Instead, I am searching for something "behind" these personal details—something unique that does not and cannot belong to anyone else. And yet, we cannot put our fingers on what this "I" or "me" stands for. It is like my shadow—always a pace ahead of me. The shadow of oneself will never wait to be jumped on; it evades capture, and yet is never very far ahead. That's why the soul may seem so elusive. Yet, it's a fact hard to deny.

When I reflect on myself, it is always "I" that is doing this. That in itself is quite amazing. Imagine a missile being its own target, or an index finger pointing at itself. Yet, "I" is capable of doing that. Self-reflection is arguably the most enigmatic part of "I." We have a fascinating situation here: I-as-a-subject—let's call it "I-now"—can reflect on I-as-an-object—let's call it "I-past." As a subject, I may interrogate I-as-an-object and then realize, for instance, that I-as-an-object made a mistake. It is here that we come in touch with this astonishing capability of my "self"—the faculty of reflecting on my "self." I may notice, for instance, that I am clumsy; I may even notice that I am laughing at myself for being clumsy; and then I may decide to tell others that I noticed how I was laughing at my own clumsiness. This is in fact an iterative process of self-reflection, making "I" act like my own shadow—"I-now" reflecting on "I-past." I can never get away from my "shadow" in the same way as I can get away from someone else's shadow.

Because I-now is always a pace ahead of I-past, I-as-a-subject is never open to investigation, let alone scientific investigation. Put differently, "I-as-a-subject" is not an object like other objects in this world, but it is their very origin, for without it, there would be no objects of knowledge comprehensible to the human mind. Whenever I reflect on myself as an object, there is someone doing this as a subject—it is as if another "I" is "looking over my shoulder." No wonder then, the human soul appears to be an elusive entity in this universe—immaterial and spiritual.

S: How can "I-now" ever reflect on "I-past," as you suggest? That sounds like pure magic to me. You make it look like "I" is able to pull itself up by its own bootstraps. I am sure this goes against the discoveries of science, in particular neuroscience!

R: Indeed, here we have another enigmatic part of the beautiful soul and mind of a human being—its capacity of *transcending* itself, also called self-transcendence. Yet, this phenomenon is something hard to deny. When I say or think "I am only human," I am not comparing myself with something "below" me (such as a cat, a dog, or an ape), but I am comparing myself with something—or rather Someone—who is "above" me and transcending me. When I call myself "only human," I am actually comparing myself with Someone who does not have the limitations I experience myself. In some mysterious way, I am reaching out into the realm of the Absolute, far beyond myself. In so doing, the "finite" catches a glimpse of the Infinite. In the words of Vatican II (*Gaudium et Spes*, 15), "Man judges rightly that by his intellect he surpasses the material universe, for he shares in the light of the divine mind." Only because we do share in the light of the Creator's Mind, we can become "creators" ourselves, capable of designing structures based on art, architecture, engineering, and scientific research.

Somehow this means that we can transcend ourselves by going beyond the person we are right now. That must be quite puzzling, though, for many people, especially those with a scientific outlook. Of course, I cannot transcend myself on my own, for that would indeed amount to pulling myself up by my own bootstraps, as you put it. Here is what St. Bonaventure astutely remarked about this, "But we cannot rise above ourselves unless a higher power lifts us up."

Therefore, we do need a higher power to lift us up, so there *must* be a higher power to lift us up. The mind's amazing ability to reach beyond its own body, or even beyond the universe, is only possible if the human mind is a "take-off" of the Creator's Mind. Because I myself was made in the image of God, I perceive more than myself whenever I perceive myself completely. I actually "perceive" God.

Since we were created in God's image and likeness, we have been endowed with a human soul, which fact sets us apart from the rest of the world. It is the soul and mind, not the body, that are made in the image of God. That is where *Homo sapiens* becomes *Homo religious*—a being who is in a bond with God. No wonder, atheists and many scientists are out to destroy, or at least deny, this human image, because it is a divine image, a reflection of God's image. They keep coming up with alternative solutions. They say, for instance, that "I" or "me" with its miraculous powers doesn't in fact exist, and then declare it merely a collective word for a collection of cells bound together by a body. But that amounts to stating that a college, for instance, does not exist, because the word "college" is just a collective name for a collection of buildings and departments. Of course, that college does exist, otherwise those buildings and departments would not even be there either.

C. S. Lewis describes the creation of the first human being as follows: "Then, in the fulness of time, God caused to descend upon this organism… a new kind of consciousness which could say 'I' and 'me,' which could look upon itself as an object, which knew God, which could make judgments of truth, beauty, and goodness." This is the birth of the human soul, which transcends the human body.

S: This all may sound nice from a philosophical perspective, but I would prefer to think about it in scientific terms. I would say that a person's identity is based on one's DNA. A person's DNA basically stays the same despite changes in all the parts of a body.

R: I am sure scientists prefer this material alternative over a non-material soul. But the question is: do they have a point? I have my doubts. My main reason is that DNA may appear as a monolithic bloc, but as we found out, it floats in the air without an environment to work in. Let's say for now, DNA must work in the surroundings of a body outfitted with a soul. As a matter of fact, cells can do different things in different parts of a person's body based on the same DNA. Somehow, then, there is more to DNA than its molecular structure. There is "something" that activates different DNA parts in different types of cells. Is this also determined by DNA? If it is, then we do not quite know how it is being done. But, yes, that may just be a temporary problem for science, soon to be solved.

On the other hand, there must be something that activates DNA. As I said earlier, DNA on its own cannot make itself work the way it does work—that would lead us into a circular form of reasoning by assuming already what one attempts to prove. DNA itself cannot make itself work the way it does, but it must instead be done by the soul. It's like with a computer program that cannot start itself, unless someone turns the computer on to begin with.

In addition, there is a more serious problem for the "identity role" you like to assign to DNA. It is currently estimated that almost 99.9% of the human DNA is similar in all human beings. This leaves us with only 0.5% of the DNA to determine the uniqueness of an individual. Of course, that does not make it impossible, but highly unlikely. Think of two identical twins. They are supposed to have the same DNA. True, mutations may interfere and thus create a few new differences in their DNA. But would a small change in DNA cause a different identity? Or reversed, even if their DNA is still the same, do they then also have the same identity? I don't think anyone would believe that. The "I" of one twin is different from the "I" of the other twin. Identical twins just do not have the same identity, even if their DNA were the same. I would say the reason for this is simple: it is not one's DNA but one's soul that determines a person's identity. Identical twins may not differ much from each other in appearance or in their DNA, but what they do differ in is the soul each one has, which makes even "identical" twins far from identical, but unique.

Let me say it once more: it is one's soul that stays the same while a person's body cells keep changing. Think of this analogy. A butterfly goes through dramatic changes in life: from egg to caterpillar, to pupa, and finally to a mature butterfly. Yet, during this entire process, the butterfly remains the same entity—in more technical terms, a member of the same species, all the way along. It may go through various appearances, but it remains the same being. In a similar way, human beings go through similar changes—from embryo to baby to toddler to adolescent to adult. These are all steps in one long, continuous growth process that starts with a fertilized egg cell and ends with natural death. To say that DNA is the only source of all of this is a moot point, so we found out. The soul must also have something to do with it.

S: Wait a minute. That goes too fast for me. I rather agree with the philosopher John Locke who says that we have no way of ever knowing the nature of anything, but only the impression it makes on our senses.

R: Well, I would argue Locke is deadly wrong on that. As a matter of fact, we do know the *nature* of things. Locke thinks that a tadpole, for instance, is not a frog but "turns into" a frog with a different appearance. He would say that a tadpole is not a frog, because it does not look like a frog. Instead, I would argue that there are real distinctions in the nature of things. To be a frog, for instance, is not to be a butterfly. What makes them different? Not the way they appear to us, but the way their "nature" makes them different. A tadpole will never turn into a butterfly, for instance. That's where the Aquinian distinction between form and matter comes in again.

We are in fact surrounded by substances that change their appearances. Most trees go through dramatic changes of appearance during their lifetime, but even during the cycle of each year. Apparently, we have a case here of accidental changes—one and the same subject may go through various changes while remaining the same entity. A thing is what it *is*, not what it *looks* like. It is not until the tree dies that a substantial change is taking place, because that very event removes the subject of accidental changes itself. Something similar holds for changes in a human being during the process from conception to natural death.

In contrast, the appearances Locke refers to are very deceiving. Rain, mist, fog, steam, ice, hail, snow, and rain have different appearances, yet they are different states of the same substance, water. Something similar can be said about human beings. They may have quite different appearances during their lifetime, but they have the same identity and have the same "self"—a human soul, that is. We are dealing here with a *human* being from the very beginning of its life—though not complete yet, but definitely in the making. As Aristotle said, we judge the nature of a horse by a fully grown horse, not by a newborn colt.

Is the Body Something We Own?

S: What you have said so far makes me wonder what then our relationship is to our bodies. If the soul is so important in your view, then the soul must be in control of the body, so the soul must "own" the body in some way and "steer" it in its own way.

R: In a sense, the body is indeed something we *have*, but it is more accurate to say that the body is part of what and who we *are*. Just as the soul is not something that we have, but something that we are, so is the body not something that we have, but something that we are. In other words, we do not "own" our bodies. In this context, Pope John Paul II has drawn attention to the two radically different meanings of the word "my." When I say, "This is my cell phone," I mean that I *own* the phone. On the other hand, when I say, "This is my wife," it should be clear that I am not claiming that I own her, but only that she is "mine" in another way.

So, what are we to make then of someone saying, "This is *my* body, therefore I can do with it as I please"? We are so hung up on possessions that we view even personhood in terms of ownership, giving us ownership even of our bodies. However, if a human being consists of both soul and body, then I can no longer say that a person "has" a body. If you still prefer to use ownership terminology and want to say you "have" a body, then you must add that you also "have" a soul with a mind and a will—in fact, a beautiful mind and soul.

On the other hand, I can still say that I "have" arms or feet or whatever. I "own" them—they are mine and not yours—in the sense that I need to give my permission if someone else would like to have one of my organs for transplantation. Yet, it is strange for me to say, "I have or own an arm." Only a person with an arm prothesis could properly say that. The "ownership" idea of a body, or any of its organs, remains odd. True, if the human body were merely a machine—without a soul, mind, or spirit—then we could alter the body in surgery and "fix" it as we would fix a car or an appliance. Once we consider the human body as belonging to a person who "owns" it as property, the human body has become an object—a thing that is owned, like we can own a machine.

You might think that "owning your body" is just a harmless, innocent issue with no further consequences. Doesn't it merely mean that the body I have is "my" body and not someone else's body? What is wrong with that? If that's all there is to it, then there would hardly be a problem. But that is seldom the case. There is much more to it. Not only does the ownership idea disconnect the body from the soul, but by doing so it also changes the body into a machine that I can use in any way I like. To put it briefly, the body is "my" body but not "mine." However, if there is indeed someone who does own my body, it is the one who made it—I don't mean my parents, but ultimately the Creator of all there is, God. Therefore, I am not allowed to destroy my own body, or any human body for that matter—for that would be a decision beyond my authority.

If the human body were indeed a machine, and not the "embodiment" of a human person with a soul, many things would be possible. For example, we could terminate it, as in suicide, or we could rent it out, as in prostitution, or we could claim it, as in rape, or we could exploit it, as in human trafficking. Would you want to accept those consequences? If that's not problematic enough for you to be very cautious about the ownership idea concerning the human body, we should address two more, quite crucial consequences of this seemingly harmless idea. One comes up in the euthanasia debate; the other one in the abortion debate.

S: I do not see the connection you are suggesting. Euthanasia is merely a form of mercy-killing. Isn't it more merciful to end the suffering of terminally ill patients than to keep them alive? And abortion is merely the removal of a piece of tissue from the mother's womb—similar to the way we can remove an appendix.

R: I can understand why you do not see the connection. The defenders of euthanasia and abortion have done a thorough job. They have changed the terminology of the discussion about euthanasia and abortion by cleverly swapping terms that reveal the truth with terms that conceal the truth.

In the euthanasia debate, they rather not speak of euthanasia, but use instead the euphemism of "mercy killing," and they defend this with a reference to a person's "right to die." It is interesting to see how they have shifted arguments and terms supporting euthanasia from "pain relief" to "mercy-killing," next to "assisted suicide," and finally to "autonomy" and "self-determination"—which means that all people should be allowed to decide for themselves whether and when they want to end their lives, without any interference from churches, governments, or any other institutions. Isn't it "their" body, a body they own?

The truth of the matter is that these terms are more concealing than revealing. Not only is the term "doctor-assisted suicide" confusing—how can it be suicide if it is assisted—it is also misleading, because the only "assistance" the patient receives from the doctor is a prescription to be filled at a pharmacy. Therefore, a more realistic and accurate description of "doctor-assisted suicide" would be "doctor-prescribed death." It may sound harsher, but it is more accurate.

No matter which terms we use for euthanasia, the bottom-line is that we do not own our bodies and therefore cannot do with them whatever we want. Euthanasia is not just discarding a machine that does not work properly anymore. It is not even a machine that I own. The value of human lives comes ultimately from their souls rather than their bodies. All bodies come with a beautiful soul and mind. Therefore, I cannot sell, let alone kill my own body. As the Catechism (2324) puts it, "Intentional euthanasia, whatever its forms or motives, is murder. It is gravely contrary to the dignity of the human person and to the respect due to the living God, his Creator."

Abortion is based on the same erroneous mantra: "It is *my* body, so I can decide whatever *I* want." The common argument for this is that the body of a pregnant woman is *her* body, and therefore abortion is *her* choice. This may sound reasonable until we realize we are dealing here not with the interests of *one* person, the pregnant mother, but also with the interests of another person, the unborn baby.

Even if a woman were allowed to do with her body whatever she wants, we still must admit that the unborn baby is not *her* body. The baby is *in* her body, yes, but it is not a *part* of her body—it is a genetically distinct being with its own identity. Therefore, abortion does not only affect her own body, but also someone else's body, that of the unborn baby. Why would a pregnant woman not allow the unborn baby a right to his or her own body, when she claims this right so insistently for herself? Doesn't the unborn baby, too, have the right to say: "This is my body, and not yours!"

The new person in a mother's womb is another person that she is now "part" of and responsible for. Parents are not owners of their children, born or unborn, but rather their guardians or stewards. In contrast, to defend the right to life for the mother *only*, those who defend abortion simply redefine the unborn being out of existence. They will tell you that what a woman is pregnant with is not a human being but merely a "tissue," a "growth," a "tumor," a "clump of cells." Once the unborn child has become a "nothing" and is no longer a separate individual, abortion has simply become removal of a clump of cells without a voice, without an identity, without desires, without interests, without a soul. It is nothing-but-ery again in optima forma. From then on, abortion is something like the removal of an appendix.

Once pregnancy has been redefined into a disease, abortion has become a "health care" issue—solely for the mother, not for the child, of course. However, redefining words does not change reality. Obviously, for an abortion, there must be something in the womb. This "something" has a uniquely human genetic program and a human soul that directs its future development. We discussed already extensively that human beings have the same identity in all stages of their development. Therefore, there is no specific stage in this continuum where abortion is not ending or killing a human life. Just as tadpoles, when they become frogs, do not change identity but only appearances, so do unborn babies not change identity, but only appearances, when they develop, get born, or become adults. Just think of this powerful analogy: in order to protect the life of bald eagles, we also need to protect their eggs.

S: You make it sound like a black-and-white issue. Some would object that the unborn baby does have the right to life, but having that right does not entitle the baby in the womb to use a woman's body against her will.

R: The problem here is that a pregnant woman and her unborn baby do have conflicting interests in the abortion debate: the interests of the unborn having the desire to live versus the interests of the mother having the desire to make free choices. What does belong to a human being is the right to life at all stages. Therefore, the unborn has the right to life as much as the mother has this right—this is a God-given right, and no one should ever take it away.

The discussion of rights can get very murky. In this case, one "right" is set against the other, and one "right" may be considered superior to the other. The late philosopher Fr. Francis Canavan, SJ caricatures this debate with a dialogue between mother and her unborn baby as follows: "May I use your body, Madam?" "Yes, dear, you may." Or "No, you are not entitled to it." In this discussion, the unborn baby is seen as an "intruder" on someone else's property. It has become a debate on "ownership rights."

In this debate, the woman wins through her pro-choice rights, whereas the unborn can lose his or her pro-life rights. Interestingly enough, though, pro-choice activists are highly selective in their pro-choice positions. They are pro-choice as long as your choice matches theirs. They are pro-choice in choosing abortion, but they are usually not pro-choice in matters of gun control, the death penalty, free speech, or slavery, to name just a few. Defending abortion makes it hard not to do the same for genocide or slavery, to just name a few moral issues. But the fact remains that a pro-choice position is at best a matter of interests, never a matter of rights. No woman has the right to deny her unborn baby the right to life. The bottom-line is that human bodies are not owned, but come with a human soul, at any stage of life.

In the abortion debate, many people think of the pregnant mother and the unborn baby as two independent parties with their own independent, private rights. However, this view obscures the natural fact that they are related to each other as mother and child. Unborn babies are in the only natural place for them to be, their mother's womb. In other words, we do not have a case here of trespassing on private property.

In spite of all we have said about the unity of body and soul, the dualism of the Cartesian view keeps permeating our culture by driving a wedge between the two. We tend to separate them and set them in an antagonistic relationship to each other—a master-and-slave relationship. It has even caused some to see the body as a prison for the soul from which the soul wants to escape.

Can Souls Be in the Wrong Bodies?

If we go by the ownership ideology of the human body, which I rejected earlier, then some might come up with a related erroneous view, which says that we may even own the *wrong* body. More in specific, this idea has strongly affected the way males and females "find out" whether they have the right body for being male or female. Which has lately become known as the "gender issue." We have here another issue where the body is separated and alienated from the soul.

S: What makes you think there is an issue with a person's gender? By declaring there is such an issue, I am afraid you will use the previous kind of reasoning for the gender question as well, turning it into an ownership and identity issue. If you do, it will require much more explanation than what you have given me so far.

R: Let me explain first why there is a gender issue. The term "gender" was introduced in 1955 to distinguish it from the term "sex." (By the way, unlike in English, some languages use the term gender to mark nouns according to whether they are masculine, feminine, or neuter.) Someone's sex is biologically determined. It determines whether someone is born as a boy or a girl. A baby's sex is basically established at the time of conception. The egg cell always carries one X chromosome, but the sperm cell introduces either an X or a Y chromosome from the father into the egg cell. If the egg cell is fertilized by a Y-bearing sperm cell, the new organism is a boy (XY), but if it is fertilized is by an X-bearing sperm cell, the new baby is a girl (XX).

Apparently, the human Y chromosome carries genes with strong male-determining properties; in fact, it is the absence of a Y chromosome that determines femaleness. However, it is not just the Y chromosome itself that is the determining factor, but rather the *SRY* gene on the Y chromosome which produces a protein called "testis-determining factor" (*TDF*). The *SRY* gene acts as a master switch and is responsible for the development of an unborn baby into a male by initiating the testes development (whereas other genes on the Y chromosome are mainly important for male fertility). When this *SRY* gene is present, early embryonic testes develop around the 10th week of pregnancy. In the absence of the *SRY* gene, ovaries develop instead.

In general, one could say that sex determination of a male depends on the testes, and in turn, testis differentiation depends on the Y chromosome and on its *SRY* gene which produces the *TDF* protein. It is indeed true that sex determination in rare cases can lead to abnormalities. However, it needs to be stressed that the existence of abnormalities does not make them "normal," as little as neurological abnormalities become "normal" by their mere existence—they still deviate from normalcy. Talking in terms of abnormalities only confuses the discussion.

Typically, after fertilization, development is "steered" in one of two directions, either male or female. We use here so-called *fetal* ages, counted from the estimated date of conception (some use *gestational* ages instead, which only start counting at the first day of a woman's last period—that's 2 weeks after ovulation). Using fetal ages, we find at the 10th week that, the penis of the male is slightly larger than the clitoris of the female. At the 12th week, the male scrotum has formed from the tissue that becomes the labia major in the female. Finally, at the 34th week, the distinctive features of the genitalia of the two sexes are fully present. Therefore, at birth, the anatomical differences between male and female are practically unmistakable. When the baby is born, the obstetrician or midwife announces, "It's a boy" (M) or "It's a girl" (F)—and it will stay that way.

But once the newborn has been "declared" a boy or a girl on a traditional birth certificate, based on anatomical and sexual characteristics, new features develop that can make them either more masculine or more feminine. During their further development, the difference between boys and girls becomes much more than a difference in biological characteristics—namely, differences in behavioral traits, social roles, and cultural expectations that come with being a man or a woman in a given society. Typically, once parents know the sex of their child, they treat the child either as a boy or a girl (unless they settle on what they had hoped for). In other words, early on in human development, parents as well as society take on a "molding" role.

This latter aspect of development is usually captured with the term *gender*. Whereas sex is engrained in the chromosomes, gender is acquired, partly based on cultural expectations, partly on personal choices. To put it simply, sex is what you are biologically; gender is what you become socially. In other words, the terms "male" and "female" are sex categories, whereas the labels "masculine" and "feminine" are gender categories. Your gender may be partially "self-made," but your sex is not. This distinction between sex and gender is important to keep in mind. For example, when we talk about homosexuality, we are talking in terms of gender. Homosexuality is not based on a change in sex—of producing either egg cells or sperm cells—but is connected with gender instead.

S: You make quite a distinction between sex and gender. I think many people believe they almost seamlessly merge into each other.

R: Yes, there has been a strong tendency, especially recently, to make them almost identical. I maintain, though, that a person's sex is a biological issue and their gender a sociological issue. However, that distinction has become very blurred in the minds of many nowadays. According to this new outlook, sex is no longer a biological concept, but it has become a social construct, in the same way as gender is. People who defend this view have chosen to change their terminology. For instance, in 2013 the American Psychiatric Association (APA) changed the term "gender identity disorder" by replacing it with the label "gender dystrophia." It is no longer seen as a disorder but as a case of distress.

This is not an uncommon strategy. For instance, until 1974, the American Psychiatric Association had considered homosexuality a mental illness or disorder. When, in 1970, gay activists protested outside the APA convention in San Francisco, the APA recommended that homosexuality be declared "normal," although only 58% of the members who voted favored the change. What is noteworthy about this is that there was no new fact or set of facts that motivated this major change; instead, it was the power of the gay lobby that initiated the revision. The psychologist Philip Hickey worded it this way, "So all the people who had this terrible 'illness' were 'cured' overnight—by a vote!" Talking about science....

As to the sex-gender issue, the problem remains that a person whose biological identity is male cannot have a female gender identity; if he think he does, it is only in his head as an imitation of the other sex. Sexual orientation is not a social construction but a biological one that is essential to our identity. That is an undeniable biological fact, which is an essential part of one's sexual identity. It is worth noting that when two people with same-sex attraction live together, one of them usually plays the male role while the other plays the female role—but these roles are obviously social roles, not sexual roles, or perhaps gender roles at best. As said earlier, appearances can be very deceiving.

However, the fact persistently remains that gender does not and cannot replace or alter sex. To think differently destroys a person's identity as a man or a woman. It obscures the reality of sex differences, making us believe that we can manipulate sex differences entirely to our own liking. It gives us an alibi to reject the binary division of persons into two sexes, so that we can claim the freedom to be either, both, or neither one, depending on our mood.

As a matter of fact, a person's body is a fundamental indication of what sex he or she belongs to. It is a physical, empirically verifiable reality that does not change simply because our beliefs or desires do. One can surgically change the genitalia and gonads one was born with, but not one's sex. One cannot "re-invent" oneself that way, because one never "invented" oneself to begin with. We must conclude from this that gender does not replace sex, nor does it nullify sex. Gender is a social construct placed on a biological foundation. Those who obscure the difference between sex and gender—a biological concept, sex, versus a social construct, gender—do so by equating both of them to a merely social construct.

In 2012, Pope Benedict XVI connected an extreme version of the gender ideology with the words of the late French philosopher Simone de Beauvoir: "one is not born a woman, one becomes so"—which denies the fact that one is born as a woman, but may not think so. Taken this way, the pontiff declared the "gender" ideology a new philosophy of sexuality: "According to this philosophy, sex is no longer a given element of nature, that man has to accept and personally make sense of: it is a social role that we choose for ourselves, while in the past it was chosen for us by society." The pontiff was basically asserting that this version of gender theory confuses sex with gender, and thus confuses sexual identity with gender identity.

S: The fact remains, though, that some people claim they were born in the wrong body. This means the sex of their body does not match their gender identity. Isn't that a fair assessment?

R: This idea is basically rooted in various strange kinds of philosophies—all of which are of a dualistic nature. We find this dualism already in Plato's philosophy where he speaks of two opposing forces working within humanity: flesh versus spirit. In Plato's world, the spirit represents everything good and wholesome, while the flesh is everything evil and degenerative. The ideology of Gnosticism—a collection of religious ideas and systems which infected Christianity as early as the 2nd century, yet is still alive and even popular today—took on this view and made it look like souls are trapped in dungeon-like bodies and therefore want to be released from their prison. All Gnostics, ancient and modern, believe that the soul is locked up in the body and needs to be freed from this dungeon. Then there is this other dualistic view that goes back to René Descartes, who not only tended to separate body and soul, but also set them against each other in an antagonistic relationship—a "master/slave" relationship, so to speak. Consequently, the body is again seen as a prison from which the soul wants to escape.

The outcome of these dualistic ideas for the gender debate is almost predictable. In these views and ideologies, an entirely isolated soul can decide on its own what the body should be like—for instance, male or female. In other words, there is a disconnect between the sex of the body and the gender of the soul. However, what is lost in this approach is the fact that body and soul form a unity, with the soul expressing itself through the body. It is not so that the soul is at the mercy of the body, neither is the body ruled by a tyrannical soul. We are ensouled bodies just as truly as we are embodied souls. In other words, there is no such thing as a disembodied soul, nor is there a soul-less human body.

How then could some say that they have a soul imprisoned in the wrong body, and therefore need sex-change surgery? As the Catholic theologian Randall Smith put it, "Only a man missing a foot who accidentally picked up the wrong artificial foot at the hospital can truly say: 'I have the wrong foot.' But a man cannot look at his own living foot and say: 'I have the wrong foot.' Still less can he look at his own body in the mirror and say: 'I have the wrong body.'" This is certainly also a serious problem for the ideology of genderism.

Nevertheless, the genderism ideology keeps claiming that we can cross the boundaries between the two sexes whenever we opt for "reconstructing" the body. But this denies the fact that there is at least one boundary that cannot be crossed: the child-bearing capacity of females. No "constructing" there (at least not for now)! Our bodies contain a self-sufficient digestive or respiratory system, but it only contains *half* of a reproductive system, and thus must be paired with the other "half-system" belonging to a person of the opposite sex in order to carry out its function. In other words, gender may be pliable, but sex is not.

An added problem of the gender ideology is that it tries to replace reality with a make-believe version of it. However, perception does not change reality. Think of this comparison. The situation of trans-*sexual* individuals comes close to the position of trans-*racial* individuals. There are many similarities. If you can be transgender, why can't you be transracial, too? In either case, there is a *decision* to change-over. We had a recent case of someone who claimed a "transracial identity," in a way similar to claims of a transsexual identity. This sparked a conversation that we did not even know we needed to have. Is it really possible for an individual to be born "in the wrong skin"? There is that soul-body dualism again. Even if you identify closely with the

Black community, that does not make you a person of African descent.

Just as transgendered men do not become women, or vice versa—for our sex is part of who we are—in a similar way do transracial individuals not become members of another race by perception, because their skin color, too, is a part of who they are. A woman declaring that she is a man is just as odd as a white woman declaring that she is "really" black. The irony is that the concepts of "transgender" and "intersex" themselves actually presuppose the very sexual difference that they propose to negate or replace.

This reminds me of what happened in the Soviet Union during the1920s, when the botanist Trofim Lysenko asserted that rye could become wheat, that wheat, in turn, could become barley, and weeds could become grains. This idea won the Communist Party's approval, but the result was a devastating famine throughout the 1930s. We cannot just deny or ignore reality if we wish to.

S: What would you say then to people who do ask for sex-change surgery? Isn't that a good way to change one's nature and one's identity with the help of science?

R: Receiving hormones of the opposite sex or having sex organs surgically removed, changed, or replaced are not sufficient to change one's sex. They may change appearances, but not the identity of a person. Nevertheless, some people have always tried to do just that. In the past, some used cross-dressing, but that has become passé with the advent of sex-change surgery. Now former cross-dressers can truly transform their appearances to an extent that clothes, wigs, and makeup never could achieve. But the idea behind all these trials is the same: changing appearances, and appearances only.

In 1979, after commissioning a study of the outcomes of sex-change operations, Dr. Paul McHugh in his capacity as chair of the Department of Psychiatry, put a halt to transsexual surgery at Johns Hopkins Hospital. He wrote, "We psychiatrists, I thought, would do better to concentrate on trying to fix their minds and not their genitalia." McHugh compares medical treatment of patients who have a confused gender identity to treating anorexia with liposuction. He calls transgendered individuals "feminized men or masculinized women, counterfeits or impersonators of the sex with which they 'identify.'"

A 2016 report by Dr. Lawrence Mayer and Dr. Paul McHugh, based on nearly 200 peer-reviewed studies of sexual orientation and gender identity, discloses some striking information. First, only a minority of children who express gender-atypical thoughts or behavior will continue to do so further into adolescence or adulthood. Second, among transgender individuals, 41% have attempted suicide, whereas only 4.6% of the overall U.S. population reports "a lifetime suicide attempt." The hypothesis of "social stress" as an explanation has so far not been corroborated. Third, one hospital's practice of surgically removing the poorly-developed genitalia of male infants and giving them female genitalia showed that, years later, most of the subjects still identified as male, although their parents had been directed to raise the boys as girls.

What makes the doctrine or ideology of genderism even more dangerous is that it affects adolescents during the time they are still very malleable, susceptible, and impressionable. Dr. Paul Hruz, a pediatric endocrinologist, says about them, "Young people constitute a singularly vulnerable group and experience high rates of depression, self-harm, and even suicide. Moreover, children are not fully capable of understanding what it means to be a man or a woman."

In the words of the Catechism (2333), "Everyone, man and woman, should acknowledge and accept his sexual identity." So, we need to unite body and soul again by being happy in our bodies, in defiance of any dualistic ideology. The Catechism (2393) states also, "By creating the human being man and woman, God gives personal dignity equally to the one and the other. Masculinity and femininity are complementary—different but equal expressions of what it is to be human. Each of them, man and woman, should acknowledge and accept his sexual identity." It is interesting to note that it was through each other that Adam and Eve discovered their own identity, in body and soul.

The late Fordham University philosopher Francis Canavan, SJ summarized this well with his characteristic clarity:

Men and women are what we become if we develop our humanity as we should. There are no people who are simply human beings devoid of sex, neither are there persons whose sex is irrelevant to their personality. Sex is not a characteristic added on to human nature; it is a constituent part of our nature.

In other words, we should be proud of who we are, not of who we *think* we are.

CHAPTER 5

Can Science Eliminate the Mind?

Now that we know how science cuts everything into smaller elements, separating them from where they come from, we can understand how this procedure isolates those smaller parts and thus affects their operation. As we discussed earlier, DNA becomes just a lifeless molecule when we separate it from the cells where it came from. Or take this other example: a cell without the surroundings of other cells in a body does no longer function properly. My claim is that the same can be said about a body separated from its soul: it becomes a mere corpse. Obviously, a corpse does not have a soul.

I hope I have convinced you by now that a body does have a soul. But how does this insight translate into the body-mind issue, you might ask. Can we infer from this that a body has a *mind* as well? That's the million-dollar question we must address next.

In Search of a Mind

S: Even if you are right that science may not cover all there is, you still cannot deny that science, especially neuroscience, has given a final blow to what some call the "mind."

Learning more and more about the *brain*, science is gradually wiping out the *mind*, and it keeps doing so until there is nothing left anymore of what people used to call the "mind."

R: As said before, if there is a soul, there is also an intellectual part of the soul, called the mind. If there is no soul, then there is no mind. But if there is a soul, as we argued there is, then there must also be a mind. Therefore, it is worth to keep searching for a mind as the intellectual part of the soul. Only if we cannot find a mind can science claim victory. But neuroscience has not gotten that far yet, and I think it will never get that far.

I do agree with you that science can study the human *brain*, but I doubt whether it could ever study the human *mind*. Science can study the human brain, as a material entity, but it cannot really study the human mind, because the mind is a non-material entity. Even when scientists pretend that they do have the capacity to study the human mind, they enter a mind-boggling situation. Why is that? To study the human mind, science needs to start from the very minds of those scientists. If this is true, then science would never be able to fully comprehend the human mind, because science itself depends on the working of the human mind. Scientists may be able to comprehend the brain, but not the mind, for they need a mind to begin with in order to study the brain.

We have a simple truth here: the brain cannot be studied without the minds of neuroscientists.

You will probably attack what I just said. You may think that neuroscience is not done by the minds of neuroscientists but by their very brains. But then I ask you: how can brains ever study brains? That would be pure magic. The physical world can never be studied by something purely physical, any more than neurons could ever discover and study themselves—imagine, neurons studying neurons! That would be like saying that DNA on its own can discover DNA. Instead, I would argue that it is only thanks to their minds that neuroscientists have made great discoveries. They use their brains to do so, but only as a tool, not as the source of their discoveries. My point is that a mind is needed to understand the brain. Brains do not understand anything, let alone themselves—only minds do. Understanding is the result of an intellectual, and therefore immaterial, activity that cannot be achieved by any material organ, not even the brain.

My reason for making this claim is that the law of gravity, for instance, did not come from Newton's brain, but from his mind. This means that the brain could never study itself—only the mind can. Put in more general terms, the *knowing* subject must be more than the *known* object, for a mind is needed to understand the brain and a subject is needed to study any object. To explain the mind in terms of physics obscures the fact that one would still need to have a mind first before one could even have physics.

I realize this may be hard for you to accept. This may sound to you as philosophical hairsplitting. Yet, I invite you to challenge me when I express my doubts as to whether neuroscientists have wiped out the human mind. Just tell me where you think they are making advances in eliminating the mind and I will show you why their claims cannot be true.

S: Well, I am happy to take on your challenge. Here is one of the advances neuroscientists have made. Scanning techniques such as functional magnetic resonance imaging (fMRI) can measure brain activity by detecting associated changes in blood flow. This technique has unlocked the brain for us as the center of various mental activities. Many neuroscientists think that through their discoveries about the human brain, they have unraveled the mystery of the human mind as well.

R: When neuroscientists claim that certain phenomena of the mind are associated with certain phenomena in the brain, they cannot conclude from this that these mental phenomena were *caused* by neural phenomena. One of the first things you learn in a science class is that correlation doesn't imply causation. I think a good scientific example of this important difference is the strong correlation Louis Pasteur found between the number of microorganisms in his flasks and the rate of fermentation. He realized that correlation does not mean causation. But even if it does, it is not clear what is the cause and what is the effect. He had to prove that the microorganisms are not the effect but the cause of fermentation by showing that fermentation does not occur when the entry of microorganism is prevented.

Just as the rooster's crow does not *cause* the sun to rise, so does the fact that regions light up on an fMRI not explain whether this lit-up state indicates they are causing a certain mental state, rather than just reflecting it. Therefore, we need to find out at least whether certain mental phenomena always correlate with certain neural phenomena; and furthermore, whether they are proportional to the intensity of the mental phenomena. If such is not the case—and mounting evidence indicates it is not—then the assertion that activities of the mind can be reduced or equated to activities in the brain has in fact been falsified. Brain activity may be a necessary condition for mental activity, but it does not follow it's also a sufficient condition.

In response to a particular thought that someone entertains, certain brain areas may very well become more active, but this does not necessarily mean that those areas *produce* the thought—the best we can say is that they *carry* the thought. Compare this to the way a radio or TV set carries news reports. The news report heard on a radio station does not come from a receiver through its antenna, transistors, and speakers. These do indeed help transmit the news, but they are certainly not the origin of the news and do not determine the content of the news they broadcast. When the receiver breaks down, the news broadcast is still there. But do not think now that it is the receiver that produced those reports—some human mind did. Let me put it very crudely: when I bash my radio, the news report may stop, but this does not mean the news was created by the radio. The radio station did not send out a defective signal—it was only the news vehicle that broke down.

S: Neuroscience has shown us that muscle groups in all parts of the body are controlled by the motor cortex located in the frontal lobe of the brain. The cortex of the left hemisphere controls muscle groups in the right side of the body; the cortex of the right hemisphere does so for the left body part. Even when we think we are in control of those muscles, it is actually the cortex that regulates what we do—not our minds.

R: One of the pioneers in neurosurgery, Wilder Penfield, made a compelling case against your argument. He showed the difference between mental events controlled by the *mind* and neural events controlled by the *brain* when, during open-brain surgery, he asked one of his patients to try to resist the movement of the left arm, which Penfield was about to make move by stimulating the motor cortex in the right hemisphere of the patient's brain. When the patient was still able to resist the movement of his left arm, Penfield deduced from this, "Behind the brain action of one hemisphere was the patient's mind. Behind the action of the other hemisphere was the electrode." In other words, one action had a physical, neural cause, whereas the other action had a mental cause originating in the mind. Therefore, he concluded the physical cause and the mental cause had a different origin and were of a different nature.

The neurophysiologist Sir John Eccles concluded from experiments like Penfield's, "[V]oluntary movements can be freely initiated independently of any determining influences within the neuronal machinery of the brain itself." This observation seems to call for a mind in addition to the brain. The cognitive scientist Jerry Fodor put it most vividly and dramatically: "If it isn't literally true that my wanting is causally responsible for my reaching, and my itching is causally responsible for my scratching, and my believing is causally responsible for my saying… if none of that is literally true, then practically everything I believe about anything is false and it's the end of the world." Do we want to pay that price?

In response to questions like these, one could argue that the brain is as much or as little responsible for thinking as the hand is for grasping or the leg for walking. In fact, the hand is not the agent of grasping, but merely an organ or tool of grasping. In a similar way, the brain can act as a tool for the mind, without being identical to the mind. The mind's *dependence* on the brain is one thing, but *identity* between mind and brain quite another.

Think of this case. When counting things, we can use our fingers or our eyes, but no matter which organ we use, it is not that organ which does the counting—not even the brain. All these organs are merely tools, whereas it is the mind that does the counting.

S: Nevertheless, the fact remains that neuroscientists can prove brain activity of the brain with brain scans, but they cannot prove activity of the mind. I conclude from this that your so-called mind activity is merely an illusion we have.

R: Those who say something to this effect are in fact proclaiming that the mental state of the *mind* is nothing but the neural state of the *brain*—which is a clear case of the nothing-but-ery or reductionism. In other words, they claim a brain scan can presumably tell us everything about what is going on in one's mind. If so, neurologists should be able to read a person's mind! I find that hard to believe.

I think it is rather obvious that there is no such thing as mind-reading through brain scans or other techniques. Contemporary neuroimaging techniques make it possible to directly observe only the *effects* of neurological activity such as changes in intracranial blood flow. But one cannot "see" mind activity itself, only the effects of mind activity. Consequently, neuroscientists cannot just "read" a person's mind. The reason is quite simple. The mind is always the mind of some specific person. The world of the mind is a private world—it can never be in two persons. Mental events happen in my private world which I have exclusively to myself. Whereas the world of my body, including the brain, is public and accessible to others, the world of my mind is private. Even brain scans have no access to someone's private world of thoughts—all they can pick up is "brain waves," but never thoughts or intentions, as those fail to show up on pictures and scans. Even a lie detector does not register thoughts, but only emotional responses to thoughts.

The fact remains that the mind just does not go away when we try to explain it away. Picture yourself watching through a mirror how a scientist is studying your opened skull for "brain waves." The philosopher Ludwig Wittgenstein once noted astutely that the scientist is observing just *one* thing, outer brain activities, but the brain-owner is observing *two* things—that is, the outer brain activities via the mirror as well as the inner mental processes that no one else has access to. So as to make the connection between "inner" *mental* states and "outer" *neural* states, scientists would depend on information that only the "brain-owner" can provide. I would say that is another reason why we should reject the reductionism of brain-mind equalizers.

R: Perhaps one of the strongest points of neuroscience is that the brain works in the same way as a computer operates, because both use a binary code based on "ones" (1) and "zeros" (0). Neurons in the brain either do (1) or do not (0) fire an electric impulse—in the same way as transistors in the computer either do (1) or do not (0) conduct an electric current. Consequently, it looks as if the brain "thinks" like a computer "thinks." There is no mind in a computer, hence we do not need a mind in the brain either.

R: There is something wrong, though, with this analogy of a stream of binary code in the brain. True, our thoughts may have a material substrate that works with a binary code, but it would not really matter whether this material substrate works with impulses (like in the brain) or with currents (like in a computer), because such material is only the physical *carrier*, merely a vehicle that carries something else, similar to the way trains can transport people or goods. One and the same thought can be coded in Morse, Braille, hieroglyphics, impulses, or whatever code language; it does not really matter, as these codes are just physical carriers, vehicles, or tools. But thoughts themselves are not any of these.

This raises the question of what it is then that sets a thought apart from its carrier? Well, thoughts are more than a binary code; they also have sense and meaning, and that is what the binary code misses. Thoughts are *about* something else, about something mental, which is something beyond themselves. And it is this very "about-ness" that a computer lacks; anything that shows up on a computer monitor remains just an "empty" collection of "ones and zeros," until some sort of human interpretation gives sense and meaning to the code and interprets it as being *about* something.

As a matter of fact, a computer does not even have a binary code, but just two different states (on/off) that a human subject uses as a binary code (true/false) so it can carry information that makes sense to other human subjects. And let us not forget besides that computers require a human maker and would still need a human subject to give their informational output some meaning or sense. Without human subjects, computers just cannot "think." This means computers cannot explain the human mind. On the contrary, they could not exist without the human mind.

S: This does not alter the fact that, thanks to the brain of neuroscientists, neuroscience has been able to make great discoveries about the working of the brain.

R: I do not dispute what you say, but you seem to be suggesting it is the brain studying the brain all by itself. I do not think that is possible: the physical world could never be studied by something purely physical. As I said earlier, understanding something is an intellectual, and therefore immaterial, activity that cannot be achieved by any material organ, not even the brain. Indeed, the brain can certainly be studied by scientists, but it is the mind of scientists, not their brains, that does the studying. The subject that studies the brain must be "more" than the object it studies.

As said earlier, the *knowing* subject with a mind must be more than the *known* object that is being studied, for a mind is needed to understand the brain and a subject is needed for any object to be studied. To explain the mind in terms of physics obscures the fact that one would still need to have a mind first before one could even have physics and understand physics. How could a mirror ever mirror itself, or how could a missile ever target itself, or how could a finger ever point at itself, or how could a camera ever take a picture of itself? In order to do any of these, we need something that "transcends" them, something that is capable of doing something "more" than the item itself can achieve. If the mind were *nothing but* the brain, then there could not even be such a thing as neuroscience.

In addition, I would say there must be something that "reads" and "interprets" the neuron firings that take place in the neuronal network of the brain. Could we perhaps say that it is the *mind* which reads, writes, and uses the neurons in the *brain*? I would say so, for something has got to be doing this. If that is true, then we may, someday, be able to fully understand the human brain, but that does not mean we will ever fully understand the human mind. The brain in itself is merely a mind-less entity. We obviously need something "higher" than neurons—something beyond them, something that transcends them. In short, one would need a mind before one can study the brain! That's why I like to speak of a "beautiful mind."

Then there is an additional problem for those who deny all of this. If the mind were just the brain, then all our thinking would be as fragile as the material our thoughts are supposedly based on or made of. They would be sitting on a "swamp of molecules," unable to pull themselves up by their bootstraps. In other words, denying the existence of mental activities is itself a mental activity and thus leads to contradiction. If the brain were the sole origin of all we know, including what we know through science, it would act as a boomerang that comes back to its maker, knocking out the truth claims of whoever launched it.

I hope all the answers I gave you in response to your objections were convincing. You are right, the mind seems to be nowhere in the chain of all we do with our bodies, as it does not occur on the physical, scientific map of the body. And yet, it is the "soul" of it all and pervades the entire body; it is "part" of everything the body does—without being a bodily part. The mind is somehow an immaterial, nonphysical entity working in a material, physical body. What all my previous answers to your questions had in common is my main point which is that the mind can do things that no brain can do. As we said earlier, the existence of the mind is part of the existence of the soul. Both are realities we cannot deny. They are both "beautiful." It is the place where thoughts are "born."

The Cradle of Thoughts

S: Now you are using a new strategy to defend your case by introducing "thoughts." You are trying to convert thoughts into something enigmatic. Well, I have news for you: thoughts are not—even thoughts do come from the brain, not from the mind. When we are thinking, we think with our brains, not our stomach. Neuroscience has shown us that it is the brain that produces and manages our thoughts.

R: I still want to stress that we have no room left for *thoughts* if there is no mind. You make it look like the brain secretes thoughts the way the pancreas secretes insulin. But thoughts are non-material entities. The German philosopher and mathematician Gottfried Leibniz uses a striking analogy to defend his claim that thoughts are immaterial. His idea was to picture the brain so much enlarged that one could walk in it as in a mill. Inside the mill, we could see movements of several parts, but never anything like a thought. Hence thoughts and the like must be completely different from simple mechanical movements of parts and particles. His analogy is striking, even when we replace mechanical movements with chemical cascades and electrical pathways neuroscience speaks of today.

There is something very peculiar about thoughts that we do not find in material things. We mentioned this already when we argued against the ideology of materialism. Unlike material entities, thoughts are not large or small, light or heavy, hard or soft—but instead true or false, right or wrong, clear or unclear. Thoughts have no mass, no size, no color. There is no true of false in the world of molecules. Just as the brain cannot distinguish between legal and illegal narcotics, so can the brain not tell false thoughts apart from true beliefs.

In order to evaluate the outcome of neural states as true or false, for instance, we need something that is *not* neural. Unlike brain processes, which are subject to physical causation, thoughts are subject to mental causation based on reason and intellect, on laws of logic and mathematics—none of which are hardwired in the brain. Thoughts have their own existence and can become causes of their own and steer our actions. Reasoned beliefs come from the mind, not from the brain or any other organ in the body.

As said earlier, another peculiarity of thoughts is that they are *about* something—about something beyond themselves. Whereas the brain is governed by laws of physics, chemistry, and biology, thoughts are not. Thoughts have *content*. How could an assemblage of neurons—a group of material objects firing away—have any content? A mere collection of ones and zeros in the neural network is not about anything until the *mind* gives sense and meaning to the code and interprets it as being about something else. Think of what we call a picture: a picture may carry information, but the picture itself is just a collection of tiny dots; it is merely a piece of paper that makes "sense" only when the mind interprets the picture as being *about* something. The same with books: they provide lots of information for bookworms, but to real worms they have only paper to offer. They can only carry information thanks to a mind that does the interpreting.

There is at least one more peculiarity of thoughts. Whereas something like pain, for instance, can be induced in a physical way, there is no evidence that experimental stimulation of specific neuronal areas is able to produce a specific thought. The presumed jump from physical matter to "thinking matter" appears to be enormous, perhaps even unbridgeable. There must be something more to "bridge this gap— the mind, that is.

S: I still do not like the idea that we need to postulate something immaterial in addition to the material elements of a network of neurons in the brain. Occam's Razor tells us to go for simplicity instead of multiplying entities *unnecessarily*. What the brain carries are impulses, not thoughts, ideas, plans, hopes, and dreams. Thoughts are just a certain pattern of neurons firing.

R: First of all, Occam's Razor is a rather simplistic rule of thumb. It could be paraphrased as "the simplest solution is most likely the right one." But that is not always, perhaps hardly ever, the case. In genetics, for instance, the simplest solutions had to make way for much more complicated processes, including gene interactions and gene switches. Something similar happened in particle physics: the bizarre models of quantum physics—from quarks to photons to gravitons to gluons and muons—are far from simple.

Second, I am not so sure that material entities are sufficient to explain the existence of thoughts. My point is that it may in fact very well be necessary to postulate something immaterial in addition to the material entity of a network of neurons in the brain. Otherwise, we end up in a vicious circle, because as Stephen Barr puts it: "The very theory which says that theories [or thoughts] are neurons firing is itself naught but neurons firing." If materialists want to claim that materialism, the denial of anything immaterial, is true, then they should realize this very thought must be more than a certain pattern of electrical activity in their brain cells; otherwise, materialism works like a boomerang that undermines its own foundation and makes its claims self-destructive.

Third, one cannot deny the mental without affirming its existence, because denials are mental activities themselves—thoughts, that is. It makes for a paradox: if this denial is true, then it becomes false. Scholars such as J. B. S. Haldane and

C. S. Lewis have worded this paradox along the following lines: if I believe that my thoughts are the mere product of neural activities, then I have no reason to believe my thoughts are true—therefore, I have no reason to believe that my thoughts are the mere product of neural activities. As said earlier, if the mind were just the brain, then its thoughts would be sitting on a "swamp of molecules," which cannot pull themselves up by their bootstraps. This would be detrimental to their truth claims; they would be nothing more than illusions, or mere sensations at best.

Fourth, like all other sciences, neuroscience dismantles its object of investigation, ultimately down to the level of a network of neurons and neurotransmitters. This is a legitimate technique, but it does not entitle us to claim that the human mind is "nothing but" a pack of neurons. It cannot prove that thoughts are merely neuronal activities and that there is nothing else to them. The human mind is in fact beyond the reach of neuroscience and must leave the human mind untouched.

S: Even if I concede that your arguments against eliminating the mind and its thoughts are very compelling, I still do not see how thoughts can have content and can be *about* something, as you claim? Seen that way, thoughts seem very ghostlike entities to me.

R: Perhaps thoughts can be made more real and concrete, though, when we write them out or speak them out. To do so, however, we typically use sentences—for example, a sentence such as "I believe that thoughts are a certain pattern of neurons firing." Well, all languages use words to make sentences. Most words have a meaning—they "mean" something to us and to other users of the same language. There is a certain relationship between words and what they refer to, which is called their *reference*. What a word refers to depends on its meaning, though, which is determined by the *concept* behind that word.

This immediately poses the question as to what concepts are. A simple answer would be that concepts are the building blocks of sentences, and thereby also of thoughts. But this definition hides a lot of additional, necessary information. Concepts must be more than thoughts, otherwise we would not be able to communicate with each other, because I cannot read your thoughts, and you cannot read mine. St. Augustine, for instance, put it this way: there must be something "that all reasoning beings, each one using his own reason or mind, see in common."

If you still find this too abstract, think of specific concepts such as snow, cell, neuron, electron, gravity—and there are a zillion others. But what they all have in common is that concepts are always *universal*, whereas material things are always particular. Concepts do not have material properties—they have no weight, no shape, no color. Therefore, the concept of "circularity," for instance, is not circular itself; the concept of "neuron" is not a neuron itself; the concept of "electron" is not an electron itself. Thinking done with concepts requires universality, whereas sensorial information is always about particulars—about "this particular neuron," for example. This explains why an immaterial mind is required so as to provide an adequate explanation of thinking in terms of concepts. We can even conceptualize what we cannot visualize—something like a circle with four dimensions, for instance. That definitely calls for something mental, not just neural.

Add to this that it is thanks to concepts that we can see similarities not immediately visible and not directly tied to what we perceive. Everyone, even animals, can see things falling, but to perceive "gravity," one needs the concept of "gravity" in order to see what no one had been able to see before Isaac Newton. The concept of "gravity" allows us to "see," for example, the similarity between the motion of the moon and the fall of an apple. In short, concepts go far beyond what the senses provide—they transform "things" of the world into "objects" of knowledge, thus enabling us to see with our "mental eyes" what no physical eyes could ever see before. They enable the knowing subject to create known objects.

I hope it is clear by now that reducing concepts to a "product of neurons" obscures the fact that "neuron" is a concept in and of itself. Doing so would make for a pernicious vicious circle again, as said earlier: the very idea that concepts are *nothing but* neuronal firing is itself *nothing but* a firing pattern of neurons. There is no way of stepping outside this vicious circle other than acknowledging that there must be more than material entities in this world. This takes me back to what I said earlier: understanding something is an intellectual, and therefore immaterial, activity that cannot be achieved by any material organ, not even the brain. It requires thoughts and concepts.

To deny that there are immaterial entities in this world—entities such as thoughts and concepts—is basically suicidal. Let me say it again, rigid materialism is detrimental to all we claim to know, even so in science. As Stephen Barr puts it, "Just as the astrologer believes that his life is controlled by the orbits of the planets, the materialist believes that his own actions and thoughts are controlled by the orbits of the electrons in his brain." Do we seriously want to accept that? I don't think so. In contrast, I would say that we do need the mind as the cradle where all our thoughts are born.

S: By saying this, you seem to create a big divide between humans and animals. You probably assume that animals do not have a mind, and therefore have no thoughts. However, biologists have narrowed down the gap that some people thought existed between animals and humans.

R: Perhaps it may look like they did, until you study their claims a bit further. Here are my reasons for doubting they have succeeded with their reductionistic claims that humans are merely animals.

1. No matter how hard we try to make animals talk, their "talk" is forced upon them. They may use sounds for communication, but that is not the same as having a conversation. They may utter sounds, especially for courtship or intimidation, but that's not language. It is in fact striking to note how humans almost "instinctively" use language, whereas animals do not, not even apes. Apes must be *forced* to "say something," whereas children do develop this capacity naturally (otherwise we would call upon a therapist for help). In addition, what these apes "say" is not a language, not even a sign-language. It has no grammar like human language does. Even the simple "sentences" taught to apes like Washoe, Koko, and Nim lack any kind of grammar—they are something like "me Tarzan, you Jane."

Besides, the "words" that apes use do not have any meaning other than referring to concrete items—they may have a reference but no content. Their words have no meaning but are merely linked to a particular situation. The chimp Nim, for instance, used the label *apple* to refer to *anything* associated with apples. Humans, on the other hand, can use words linked to concepts rather than to particular situations. They can talk about a "mammoth" without ever having seen one. They can talk about "Atlantis" without ever having been able to locate it. They can talk about "π" without ever being able to find an example of it in the real world. They can talk about predators, without having any particular predator in mind. Animals cannot do any of this.

2. No matter how hard biologists have tried to make animals use tools, their "tool-use" is so primitive compared to ours. Chimps, for instance, have been able to "see" that a stick can help them reach food by making such a "physical" connection when they see a stick *near* food. However, when the food and the stick are kept far apart from each other, they cannot make the connection—even if they were able to do so before. Apparently, the stick is not a mental thought for them but rather a physical signal that can invoke an intelligent reaction; in other words, a chimp has a physical image of a stick in relationship to food, but it has no mental idea of what a stick is and can do in various situations. Why not? Animals are fully enveloped in their surroundings, they are caught up in a network of physical connections, unable to transcend the current situation with the mental power of abstract thinking and reasoning. In short, they lack thoughts and concepts.

3. No matter how hard we try to make animals think like we do, their "thinking" is not a matter of reasoning with the use of concepts. Animals live in a world of signals, while humans live in a rich world of signs, symbols, and concepts. When prey animals flee away from a predator, they are not reading the predator's mind but they just process certain

signals. When prey animals spot a hawk in the sky, they are not identifying the bird with a concept—like "hawk," let alone "predator"—but merely as a signal of imminent danger that may require direct action. They do not study birds in the sky like ornithologists do; they do not think in terms of true and false; they do not think in terms of cause-and-effect; they do not think with "if-and-only-if" statements. Instead, they are "moved" by motives, drives, instincts, emotions, stimuli, associations, and training, but not by reasons or mental concepts.

In other words, animals may be able to communicate with each other through signals—just listen to birds on an early spring morning. However, signals are not concepts; those birds are having an exchange of signals, but not an exchange of ideas or thoughts. Signals refer directly to a specific thing or situation, whereas concepts do not. In short, animals see only signals that call for an immediate response. What they cannot see, hear, smell, feel, or taste does not exist for them. They live in a world of signals, while humans live in a world of concepts. Signals call for direct action, whereas concepts do not.

4. No matter how hard we try to let animals have reasons for their actions like we do, their "reasons" are merely motives driven by food and sex. Animals treat everything in their surroundings as signals that call for a direct response (association), but they cannot use thoughts to ponder realities beyond their needs for food and sex. Animals drink be-

cause they are thirsty—which is not a mental reason but a material cause. Human beings, on the other hand, can drink because they have various kinds of reasons other than quenching their thirst.

What I conclude from the above arguments is that the human mind keeps standing tall as a unique entity in the world of the living which cannot be found anywhere other than in the world of human beings. In spite of neuroscience's efforts to dismantle its object of investigation as a mere network of neurons, it is hard, if not impossible, to get rid of the human mind as the cradle of all thoughts. That's what makes the mind so beautiful.

When the Mind Seems Gone

S: The mind may be quite something in your way of thinking, but as we all know, it happens sometimes that the mind seems to be gone when the brain is in trouble. I am talking about people in coma, those with brain disorders, and so on. How are you going to deal with that problem?

R: I keep stressing that the mind may operate through the brain but is not identical with it. The brain is at best the machinery of the mind, without being identical to the mind. So, the machinery may be affected, but the mind is different from the machinery it uses. Even people who have "lost their mind," as they say—through dementia, Alzheimer's disease, autism, or mental insanity—have not really lost their minds or their souls, but there are certain defects in the physical network of neurons, neurotransmitters, etc., which prevent the mind from working through this network the way it used to, or usually does. Sometimes, for instance, people cannot say what their minds want to say—because of a coma, for instance. In cases like these, the mind has become an "incarcerated" mind working with a failing brain.

As we said earlier, it is not the radio that produces the news reports; it only carries and broadcasts them. In a similar way, we should never overlook what is "behind" anyone's human body and brain, even when the mind seems incarcerated. Even then, we are still dealing with a beautiful mind and soul. The brain acts as the physical carrier for the mind's immaterial thoughts. If something is broken in the carrier, that does not mean there is also something broken in the mind. Think also of people with dyslexia or a speech impairment; it is not their minds that do not work properly, but there is something "broken" in their brains.

As William Shakespeare said in *King Lear*, "We are not ourselves when nature, being oppressed, commands the mind to suffer with the body." To be sure, thinking presupposes a functioning brain, but it cannot be reduced to this fist-sized organ. On the other hand, a malfunctioning brain would not directly and necessarily affect the mind. A broken brain is as physical as a broken bone—but the mind is not physical, nor is the soul. Even people who have "lost" their minds have not really lost their minds or their souls. What they did lose somehow is a part of their brains; they have lost part of their "communication channel." However, a breakdown of the brain is not necessarily a breakdown of the mind. To use my previous image again, the news is still there but it can no longer come in through a broken radio; there is still news, but we can no longer clearly hear or communicate it.

S: You are making quite a claim here. Do you have any evidence that your explanation is more than a private opinion?

R: I think I do. Some evidence comes from researchers at places such as the University of Virginia and the University of Vienna, Austria, who have been studying a phenomenon that is called "terminal lucidity"—the unexpected return of mental clarity shortly before the death of patients suffering from brain disorders. It is the term used when dying people, who have previously been unresponsive or minimally responsive, suddenly gain "clarity of mind" for a few hours, often talking coherently with loved ones before passing away a short time later. Examples of this phenomenon include case reports of patients suffering from tumors, strokes, Alzheimer disease, and schizophrenia.

We do know that there is no observable sudden change in the brain when death is near. Is it possible, then, that the mind's sudden and short-lived return to normalcy just before death is brought about, not by some inexplicable surge in brain activity, but by the soul and mind distancing themselves from the brain? Although terminal lucidity has been reported for around 250 years, it has received little scientific attention because of its complexity and transience—or perhaps because it goes against the now widely-held misconception that whatever the mind does is nothing more than what the brain does.

But I think there is even more evidence on this issue. In 1975, Raymond Moody, a medical doctor, published *Life after Life*, a book on "near-death experiences" of patients who had been resuscitated after clinical death, many of whom had extraordinary experiences of separating from the body and eventually returning back to life. Well, that book was just the beginning. It has been followed by a multitude of books and articles documenting thousands of cases in which resuscitated individuals reported viewing procedures in operating rooms, of passing through walls, of seeing relatives and friends in waiting rooms, of going through a tunnel toward a bright light radiating love and compassion, of often meeting deceased loved ones along the way, and then either were sent back, or chose to return, to their bodies. These are the phenomena that are usually reported after an individual has been pronounced clinically dead or was otherwise close to death—hence the term *near-death-experience* (or NDE). Those who have had such experiences usually swear by them and remain utterly convinced that there is a dimension of a person that survives physical death.

In fact, according to a recent Gallup poll, approximately eight million Americans claim to have had a near-death-experience, although the exact number of people who had these experiences may be much higher, because people with this experience may not feel comfortable discussing with others what they experienced, especially not when such an experience is often understood as a paranormal, or even weird, incident.

S: The evidence you bring up does not sound very scientific to me. I consider it still anecdotal evidence. As a scientist, I find it hard to believe that the mind can survive without a properly functioning brain. Is there anything that corroborates your claim?

R: Probably the most extensive, ground-breaking clinical study of NDE was published in 2001 by Pim van Lommel, a cardiologist in the Netherlands. With his team, he studied a group of Dutch patients who had been brain-dead from cardiac arrest but were successfully revived. Of the 344 patients who were successfully resuscitated after suffering cardiac arrest, 62 had experienced "classic" NDEs, which included out-of-body experiences. Of these 62 patients, 50% reported an awareness or sense of being dead, 24% said that they had had an "out-of-body experience," 31% recalled moving through a tunnel, whilst 32% described meeting with deceased people. None of those patients reported a distressing or frightening NDE. The patients remembered amazing details of their conditions during their cardiac arrest despite being clinically dead with a flat EEG of brain stem activity.

The cardiologist Van Lommel assumed that his findings supported the theory that consciousness had continued despite lack of neuronal activity in the brain—with a flat EEG, that is. Van Lommel poses the million-dollar question: "How could a clear consciousness outside one's body be experienced at the moment that the brain no longer functions during a period of clinical death with flat EEG?" Based on his findings, Van Lommel comes to a powerful conclusion:

Our most striking finding was that Near-Death Experiences do not have a physical or medical root. After all, 100 percent of the patients suffered a shortage of oxygen, 100 percent were given morphine-like medications, 100 percent were victims of severe stress, so those are plainly not the reasons why 18 per cent had Near-Death Experiences and 82 percent didn't. If they had been triggered by any one of those things, everyone would have had Near-Death Experiences.

What is even more remarkable is that pediatrician Melvin Morse found similar results among young children. That's quite amazing because most children never heard or even had the occasion to hear about near-death-experiences, nor are they motivated by personal, cultural, or religious beliefs. Morse compared his study group of 12 children, who were resuscitated from cardiac arrest or who had returned from deep comas, with a control group of 121 children who were severely ill but not resuscitated or in deep coma, and an additional group of 37 children who had received large doses of mind-altering drugs but were also not resuscitated or in deep coma. He found that none of the 121 children in the control group experienced anything like a near-death-experience. In the study group, on the other hand, 8 out of 12 had near-death-experiences.

Striking in all NDE cases is the accuracy with which NDE patients report their experiences. In the Dutch study, one man who had been in a deep coma, later told a nurse that he recognized her and saw where she had placed his dentures during resuscitation efforts; he even described the cart into which she had placed them. Similarly, Melvin Morse reports that a woman had knowledge of a shoe on a window ledge outside the hospital—this was not near the room where the patient was resuscitated, but next to a fifth-floor office where she was being interviewed. The psychologist who did the interview had to crawl along the ledge outside her window to verify the claim and to find the shoe indeed there. Some patients even reported leaving the operating room and visit their relatives and friends in hospital waiting rooms. One patient reported seeing her young daughter wearing mismatched plaids. Another woman overheard her brother-in-law talking to a business associate in the hospital waiting room in a very derogatory manner, and she was able to report this back to him later.

Though it is quite significant that patients with normal eyesight are able to report with great accuracy sensorial data that occurred while they were unconscious, it is even more remarkable that *blind* patients are able to do the very same thing with the same degree of accuracy. The psychologist Kenneth Ring and his coworkers studied 31 blind patients of whom 14 were blind from birth and evidently had no experience of seeing. They found that 21 of them had a near-death-experience and 10 of them had out-of-body experiences only. The fact that they were accurately reporting what they could not have seen with their physical eyes and bodies gives a high degree of credibility to a non-physical existence during a near-death-experience. Kenneth Ring summarizes his findings as follows: "80 percent of our thirty-one blind respondents claimed to be able to see during their NDEs.... [They] often told us that they could see objects and persons in the physical world, as well as features of otherworldly settings."

S: I would respond that scientists owe it to their profession to be critical about such conclusions. They should come up with scientific explanations for these phenomena. Have they ever?

R: Of course, they have. People with a critical or skeptical mind have the right, or even the duty, to question such conclusions. As to be expected, all kinds of biological explanations have been suggested instead: oxygen deprivation (anoxia), high carbon-monoxide levels, REM-sleep phenomena, psychedelic agents, hallucination. However, the question remains why not all people under those circumstances had ND-experiences.

In addition, more research has been done to rule these alternative explanations out. A 2019 study by Sam Parnia, MD from NYU Langone Medical Center suggests that NDE patients are "effectively dead," having no neural activities necessary for dreaming or hallucination. Besides, in order to rule out the possibility that near-death-experiences resulted from lack of oxygen, Parnia rigorously monitored the concentrations thereof in the patients' blood. He found that none of those who underwent the experiences had low levels of oxygen. He was also able to rule out claims that unusual combinations of drugs were to blame because the resuscitation procedure was the same in every case, regardless of whether they had a near-death-experience or not.

Moreover, the question remains of how such biological explanations can even begin to explain the fact that those who have "exited" their bodies can describe their environments with such remarkable accuracy. And how can these theories possibly explain that people, blind from birth, correctly see objects and colors in the surrounding areas of where they died?

As I said before repeatedly, science may not be able to answer all our questions, because soul and mind are not material entities. Yet, I must admit that science may help religion here to address the question as to whether the soul can be separated from the body—just as science did once before, for instance, when it claimed a beginning of the universe, which was later confirmed by the Big Bang theory.

S: But I think there is much more at stake here. Would not all of this mean that the mind could even survive death? If so, we might be entering here the issue of life-after-death, which is highly questionable in scientific terms.

R: Yes, it is a fair question as to whether the above findings have anything to do with life-after-death. As a matter of fact, NDE is often cited as evidence for the existence of the human soul, of the after-life, and perhaps even of heaven and hell—all of which are concepts well known from some religious traditions. Many individuals who experience an NDE do see it as a verification of the existence of an afterlife—including those who had agnostic or atheist inclinations before the experience.

As a theologian and philosopher, Fr. Robert Spitzer, S.J. has studied these reports, but he is very careful in his assessment of such NDE phenomena, for this is dangerous territory, hiding potentially explosive landmines. Yet, he concludes that these findings "suggest strongly that human consciousness, perception, memory and movement survive bodily death... and in many cases the surviving persons are led to a loving and heavenly domain." Is it not very likely, Fr. Spitzer speculates, that these experiences demonstrate the existence of a trans-physical dimension to the self? They are hard to explain by any other hypothesis. Yet, he feels entitled to conclude, "We have evidence that human beings survive bodily death, that we have a soul that literally leaves the body." Then he elaborates:

The corroborated veridical sensorial knowledge by both sighted and blind patients is very significant because there does not appear to be any physical explanation for these corroborated phenomena, leading to the conclusion that there must be some form of nonphysical conscious existence (including self-consciousness, memory, intelligence, and self-identity), and some survival of nonphysical embodiment (which allows for interaction with the physical world).

So, the question remains: what do these NDE findings prove? Probably not much in the technical sense—it is borderline science. It is certainly not a proof for eternal life, if only because there is no guarantee this nonphysical survival of consciousness and sensation will last forever. Besides, although these are impressive cases, with all being "near death," none of the people who had a near-death-experience actually *died*—they did come back. Yet, these findings do indicate that it is difficult, if not impossible, to explain NDE phenomena in purely bodily terms—given the fact that the brain is just not working during a flat EEG.

Perhaps we may cautiously assume that ND-experiences do offer us a peek through the "birth canal" of death—some kind of window into the after-life. Death may not be a final destination, but there may very well be a better purpose awaiting us. We should keep in mind that most neuroscientists would not allow for any non-material explanations anyway. But I like to point out to them not to let hard-core proponents of materialistic explanations talk away anything that cannot be counted, measured, or quantified—including life-after-death. Let us keep in mind that there is a real problem for those who say that physical death is the only *certainty* there is. If "physical stuff" is all that counts, then all our certainties are on their way out as well. To paraphrase what I said earlier: if I am certain that everything is physical, I would have no reason to suppose that this certainty is true—and hence I would have no reason to be certain that everything is physical, and physical only.

In brief, what these near-death-experiences do suggest is that the soul—and its intellectual part, the mind—can survive brain-death. These findings seem to at least confirm that there can be mental activities without neural activities associated with a flat-EEG. Even when people "have lost their minds," there may still very well be a mind behind the "broken" brain. In other words, my mind is the "soul" of everything I do; it is behind everything my body does. In that sense, we might say it is part of everything I do with my body, and yet it is not a physical part of my body. That's the enigma we must live with for the rest of our lives. That is why we can say we have a beautiful mind and soul! Well, let's find out next what their beauty is.

CHAPTER 6

The Beauty of the Soul

We discussed extensively what is wrong with reductionism. But that leaves an important question unanswered. If it is true that human beings are more than members of a colony, more than overvalued animals, more than bundles of instincts, more than complex machineries, more than packs of neurons, more than strings of DNA, more than mixtures of chemical compounds, or more than mostly empty space, what then is it that makes them more than any of these?

I hope and pray that I have convinced you so far that each human being has not only a body but also a soul. Once we do away with belief in a soul coming directly from God, we end up being "nothing but" biological creatures, or just "stardust," as Carl Sagan liked to say. In this latter view, we have lost the dignity of human beings, including all that comes with human dignity such as rationality and morality. By narrowing the gap between human and non-human creatures, we take the human soul out of the human body and thus take everything away that comes with the soul.

It is this very human soul that has both an *intellectual* part, the mind, and a *voluntary* pert, the human will. The intellectual part is associated with rationality, the voluntary part with morality. Now the question is: what then is it that makes these two parts so beautiful?

The Beauty of the Soul's Intellectual Part

S: When you say that the intellectual part of the soul is associated with rationality, what then do you mean by rationality? It sounds to me like a very vague concept that just obscures our discussion.

R: Rationality relates to the intellectual part of the human soul, giving us the faculty of making rational decisions. The faculty of rationality lets human beings be rational beings who use reasoning—which is expressed in the official name our species was given, *Homo sapiens*. This does not mean of course we always think rationally, but at least we can, and we should if possible. Besides, saying that rationality makes us think rationally sounds like a tautology that does not explain much. Let me try better.

The truth of the matter is that rationality is our main "tool" to get to the truth, to the world of what is true and what is false. This is done by reasoning. Sound reasoning is an important "tool" to discover the truth. If we did not have rationality, we would not even know about the existence of truths and untruths. This gives rationality an extremely specific scope: the search for truth. At the same time, it also has no bounds: it can be applied to an unlimited range of questions and situations.

As a matter of fact, each time we engage in a discussion or dispute, we use reasoning to defend our position or to explain why we disagree with the position of others. That is basically what we have been doing through this entire book. Each time we look for any kind of explanation, we are in search of some form of reasoning that has the potential power to convince others and reveal the truth.

In regular life, we use reasoning to make rational decisions. In debates, the debaters need reasoning to either support or attack the arguments brought forward by their opponents. In court, lawyers use reasoning to make their case either for or against the defendant. In science, scientists use reasoning to have their claims confirmed or refuted by their experiments. And the list goes on and on. In this book, for instance, we used reasoning and arguments for the existence of the soul, after we had used reasoning to argue that materialism is flawed and self-destructive.

It is important to stress, though, that reasoning needs a starting point in addition to a strong foundation. Knowledge without any firm foundation is a dangerous undertaking. All of us tacitly assume in our actions the objectivity of nature and the ability of the mind to grasp its truth ever more comprehensively. We said something similar already about the foundation of scientific knowledge, for science cannot take credit for its own starting point. As we discussed before, in order for science to get off the ground, it needs to make a series of assumptions that science cannot validate on its own. It must literally assume that an external world exists, that this world is both orderly and intelligible, that the human mind is reliable, and that scientific investigation, in the final analysis, will prove to be worth the effort of finding the truth.

No matter how obvious this seems, many philosophies have tried to undermine this very foundation. The rare exception to this form of skepticism is a philosophy of *common sense*. How can common sense bring the so much needed realism and objectivity back to human knowledge? Albert Einstein once begged for this: "The belief in an external world independent of the perceiving subject is the basis of all natural science." Well, it is exactly St. Thomas Aquinas' philosophy that has sometimes been described as a philosophy of common sense. G. K. Chesterton even put it this way, "The fact that Thomism is the philosophy of common sense is itself a matter of common sense." And then he adds, "To this question 'Is there anything?' St. Thomas begins by answering 'Yes'; if he began by answering 'No', it would not be the beginning, but the end." That's the point where all flawed reasoning ends, so that sound reasoning can begin.

There is no doubt that Aquinas did use sound reasoning. Chesterton says about him: "he had an intellect that could act like a huge system of artillery spread over the whole world; he had that instantaneous presence of mind in debate, which alone really deserves the name of wit. But it never occurred to him to use anything except his wits, in defense of a truth distinct from himself. It never occurred to Aquinas to use Aquinas as a weapon." Unfortunately, the plea for common sense voiced by people like Thomas Aquinas and Albert Einstein seems to have lost its common appeal nowadays.

Getting back to your original question, the technique of sound reasoning and rigorous argumentation is made possible by rationality. The power of rationality has been used throughout history in various forms of mathematical, philosophical, scientific, and religious reasoning. Let me give you a few examples of what rationality can do for us.

1. A simple example could be the kind of reasoning that Aristotle used after he had observed lunar eclipses: "[S]ince it is the interposition of the earth that makes the eclipse, the form of this line will be caused by the form of the earth's surface, which is therefore spherical." Columbus was another one who used reasoning to argue that the earth is not flat: as a ship sails away from shore, the upper portions of it remain visible to a watcher on land long after its lower parts are no longer visible—therefore the earth must be round. It is amazing what the reasoning of rationality can do for us.

2. Another interesting case of reasoning was used by the biologist Peter Medawar. Anyone looking at paintings of the famous Renaissance painter El Greco will notice that most of his figures are unnaturally tall and thin. Some scientists

came up with the hypothesis that El Greco must have had a form of astigmatism that distorted his vision and led to elongated images forming on his retina. Others tried something similar to explain Vincent Van Gogh's preference for yellow colors by hypothesizing a visual disorder called xanthopsia; some even mentioned drug use or glaucoma as the causing agent. Medawar showed us that the mere use of reason could have saved these scientists embarrassment. Even if El Greco did see the world through a distorting lens, the same distortion would apply to what he saw on his canvas. These two distortions would cancel each other out, so to speak, and the proportions in pictures would remain realistic. Even if Van Gogh did view the world through a yellow filter, he would also view the colors on his canvas that way. In either case, it is amazing what the reasoning of rationality can do for us before we dive into a needless scientific discussion.

3. Dimitri Mendeleev, who discovered the Periodic Table of Elements, had to conjecture that atomic weight was the attribute according to which the elements should be ordered. Interestingly enough, he made his discovery just a few years after the concept of atomic weight had been clarified. Yet, he needed sound reasoning next to put things together. When he attempted to classify the elements according to their chemical properties, he noticed patterns that led him to postulate his Periodic Table. Thus, he could predict certain characteristic properties of elements based on their atomic

weights. He looked at the elements not as separate, isolated components of nature, but as part of a larger reign of order with a periodic structure. His Periodic System was practically conceived from behind his desk, not in the lab. It is amazing what a rational mind can do for us.

4. The reasoning that comes with rationality can also prove that God exists. To begin with, we must accept that everything that comes into existence must have a cause, for nothing can cause itself to exist—which is a premise no one in his right mind can deny. Next, we must accept that any series of causes, even if it is an infinite series of causes, must have a cause as well, otherwise it has no foundation to rest on and merely "floats in the air." Given these two premises, the conclusion necessarily follows that there must be a First Cause, outside of all other causes, which is a necessary Being, is Existence itself, does not need a cause for itself, and makes all the other causes possible. Without this First Cause, no secondary causes could even exist. This conclusion is evident by mere reasoning and is based on the human faculty of rationality. It is amazing again what a rational mind can do for us.

5. A more general form of reasoning allows us to establish a claim by showing that the opposite scenario would lead to absurdity or contradiction, and therefore needs to be rejected. A simple example could be this: if the earth were flat, then we would find people falling off the edge; well, we haven't seen that, so the idea of a flat earth is absurd. A more

sophisticated and abbreviated example runs like this: if there is only matter in this world, then truth is something immaterial cannot exist—therefore, believing that matter is all there is cannot be true. It leads to absurd or contradictory consequences. Or take this one we used earlier as well: if things would make themselves exist, then they would have to exist before they came into existence—which is absurd and must therefore be rejected on rational grounds.

I hope these few examples are enough to show you the power and the beauty of rationality and its power of reasoning. None of these would be possible without the faculty of rationality.

S: Perhaps what you call rationality is not as amazing as you suggest. I would say, in line with many others, especially scientists, that this capacity must come from our genes, or perhaps even from our non-human ancestors.

R: First of all, animals do not have thoughts. As we discussed in an earlier section, thoughts must come from the mind, not the brain. Animals do have brains, of course, some of them have even brains very similar to ours, but they do not have minds, and therefore, they have no thoughts. Animals act based on instincts, emotions, stimuli, associations, and training, but they have no reasons or mental concepts for their actions.

Second, there is no such thing as truth in the animal world. All that matters in the animal world is fitness, not truth. To know that the earth is not flat or that the universe had a beginning are not truths that have any survival value—they do not affect fitness in the struggle for survival. Truth is something only the human mind can discover; it is a mental concept. Thanks to our minds, we know about the truth of gravity, for instance. Animals experience the effects of gravity, like we do, but they lack the concept of gravity. To take another example: if fish do not "obey" hydrodynamic laws and if birds do not "obey" aerodynamic laws, then their fitness is in jeopardy. But they do not know about such laws and do not know these laws are true. They do not know either, and do not need to know, that the pull of gravity between two objects is proportional to the masses of the objects and inversely proportional to the square of the distance between their centers of mass. Humans, on the other hand, do know about the laws of nature, but that's not something they share with the animal world—it's thanks to their intellect and rationality.

Third, in the animal world there are no concepts. As we explained earlier, concepts are general and go beyond what the world shows us through the senses. Seeing a predator or a tree is a matter of perception—something animals can do too—but using a concept is a matter of thinking and rationality. According to the philosopher Mortimer Adler, there is conceptual content and perceptual content. Conceptual content is about predators and flies in general, not about one particular predator or fly. It is universal, not particular; it is something mental, not neural, and therefore requires a mind. Thoughts and concepts have meaning and content, while perceptions do not, which is the reason why they do not exist in the animal world. Think of this example. The concept of "gravity," for example, allows us to "see" the similarity between the motion of the moon and the fall of an apple. But that is conceptual content, not perceptual, and is therefore lacking in the animal world.

Fourth, animals seem to live their lives entirely in the present, without having any thoughts about the past or the future—perhaps memories, but not thoughts, concepts, or interpretations. If pets have a pedigree, it is thanks to their owners; if they have birthdays, wish lists, appointments, or schedules, it is because their owners create those; and if they have graves, those were dug by their owners as well. What a disparity there is between them and us! Only humans are conscious of time; they can study the past, recognize the present, and anticipate the future; they even desire to transcend time, thinking about living forever. Only humans wonder "what caused or will cause what and why?" Only human beings have inquisitive minds asking questions such as "Where do we come from?" and "Why are we here?" Only humans have the capacity to be scholars and scientists; they can even study animals, whereas animals can only watch humans but never investigate them.

In other words, unlike animals, human beings are always in search of some kind of worldview or explanation of life—which certainly goes far beyond their need for food. In short, human beings are driven by rationality, which gives them the capacity to have thoughts and make rational decisions. In a nutshell, they are rational beings.

S: What you have said so far does not alter the fact that animals do show various kinds of intelligence. I would say, the human intellect and its rationality must at least be a takeoff of their intelligence.

R: I am not so sure about that. I think you are confusing intelligence with intellect. Rationality is not a matter of intelligence but of intellect. Whereas intelligence can be graded on an IQ scale, intellect cannot. One can have intelligence to different degrees, but one cannot have intellect to different degrees. Intelligence only works with perception of sense-data. Consequently, animals may show various forms of intelligence in their behavior, given the fact that intelligence is a brain feature and as such an important tool in survival. We find spatial intelligence in pigeons and bats, social intelligence in wolves and monkeys, formal intelligence in apes and dolphins, practical intelligence in rats and ravens, to name just a few forms of intelligence. Intelligence is a matter of processing sense-data in a "clever" way—something even a robot with "artificial intelligence" can do by "cleverly" or "intelligently" processing sounds, images, stimuli, signals, and the like.

Intellect is very different from this. Like intelligence, intellect also uses sense-data, but unlike intelligence, it changes perception into cognition by means of mental concepts and reasoning, which makes sensorial experiences *intelligible* for the human mind. A mental concept may be as simple as a "circle" or as complex as a "gene," but a concept definitely goes beyond what the senses provide. We do see things falling to the ground, but we do not see gravity. We do not even see circles, for a "circle" is a highly abstract, idealized concept (with a radius and a diameter). The history of science is full of examples to illustrate this. For instance, before Alexander Fleming discovered penicillin, many before him had seen molds in their bacterial cultures, but it took someone like Fleming, who had always been searching for inhibitors of bacterial growth, to notice a zone around an invading fungus where the bacteria could not seem to grow. Only then was he able to make sense of these phenomena by using the concept of what soon came to be called antibiotics.

Thanks to the human intellect with its concepts, we are also able to peer further back in time, and further forward into the future. For animals, on the other hand, there is no "tomorrow," because that requires the concept of "tomorrow." That's why animals cannot think about the possibility that someday there may not be a "tomorrow" in their lives. How different is the human species! No other species would ever think to ponder the age of the universe, or to wonder how it will end. Reasoning leads us from one idea to a related idea; it is a matter of pondering realities beyond what we experience through our physical senses, thus allowing us to transcend the current situation through the mental power of reasoning. Rationality is our gateway to truth. Animals, on the other hand, treat everything in their surroundings as signals that call for a direct response, but they cannot use concepts and reasoning to ponder realities beyond their needs for food and sex. They act mainly by "instinct," not by reasoning. Even when they use their intelligence, they do so without having intellect.

Let us go back for a moment to what St. Thomas Aquinas says about the intellect. Aquinas' famous motto is that "nothing is in the intellect that is not first in the senses." He basically distinguishes passive sensation from active perception. The senses provide the material the intellect processes. For Aquinas, the senses are only the starting point of knowledge. Hence, sense experience begins in the senses, but it is completed in the intellect. For Aquinas, sense experience provides only the "passive" component of knowledge; in addition, there is the intellect which provides the active component of knowledge. As he puts it, whereas the eye only sees, without seeing that it sees, the intellect knows what it sees, and also knows that it knows. All we know about the world does come through our physical senses but is then processed by the immaterial intellect that extracts from sensory experiences that which is intelligible for the intellect with the aid of concepts, thus transforming concrete data into abstract entities. It is obvious that animals do have the former part, passive sensation, but not active perception which is only possible through the intellect that animals lack.

In a nutshell, it is through the intellect with its concepts that humans are able to transform "things" of the world into "objects" of knowledge. It is rationality that makes the world intelligible and understandable for us; it gives us the power to comprehend what we experience. As we stressed many times before, denying the existence of the mind and its rationality is self-destructive. Isn't that a beauty to behold!

S: If our faculty of rationality does not come from our genes or from our non-human ancestors, as you claim, where else could it come from? It must then be either the result of mere coincidence or must be caused by some mysterious source.

R: Somehow the world we live in must be such that human rationality can have access to it. If the world were chaotic, our minds could not give it the order it does not have. If the world were a complete riddle, our minds could not make it intelligible. There must be some kind of independent "match" between how the world is in reality and what we know about the world through our minds. There could be no science if there is no real world to understand; neither could there be any science if there were no minds capable of grasping the real world as it is. Our intellects can only make the world intelligible on condition that the world is such that we can grasp its order with our minds. That "match" must be an enigma for most people as it seems to be for you.

This enigma has even startled hard-core scientists. The Nobel Laureate of Physics Eugene Wigner published in 1960 a paper which he gave the provocative title *The Unreasonable Effectiveness of Mathematics in the Natural Sciences.* His point was to show how difficult it is "to avoid the impression that a miracle confronts us here, quite comparable in its striking nature to the miracle that the human mind can string a thousand arguments together without getting itself into contradictions, or to the two miracles of laws of nature and of the human mind's capacity to divine them." He asked himself whether the match between mathematics and physics is merely a happy coincidence that seems to be "unreasonable" and hard to explain. Although he did not give us an answer, I would like to claim that coincidence is not really a satisfying explanation.

I think it is only in religion that we can find a solution for this enigmatic harmony or correspondence between reality in the world and the intellect of the human mind. The Judeo-Christian religion is very definite as to what unites the two: on the one hand, the world is a product of the Mind of God; on the other hand, the human mind is also a product of the Mind of God. That's why the two can "communicate" with each other. Somehow, there is reason in the human intellect, and there is reason in the world around us. That's why they can match each other so beautifully.

The order found in the universe—based on laws of nature—matches the order found through our rationality—based on intellect, logic, and reason. The mystery we have here can be best, and probably only, explained if they both have the same source—God the Creator. Only God can make it possible that the rationality present in the human intellect matches the rationality we find in the natural world. The only rational solution of this "match" is this: they both derive from the Reason we find in the Mind of God. Only this can explain the power of reason in both the physical world and the human intellect.

Pope John Paul II's encyclical *Fides et Ratio* offers a similar view when he explains how the activity of science is possible: "It is the one and the same God who establishes and guarantees the intelligibility and reasonableness of the natural order of things upon which scientists confidently depend." Undoubtedly, the conformity and harmony between reality and rationality must be a riddle for non-believers, but for religious believers it would be "a match made in heaven." The existence of a Divine Intellect is the best—and I would add only—explanation of why there seems to be such a perfect "match" between the rationality of our minds and the rationality found in the world around us. No wonder then that Fr. Georges Lemaître could describe God as, "One Who gave us the mind to understand him and to recognize a glimpse of his glory in our universe which he has so wonderfully adjusted to the mental power with which he has endowed us."

S: I can accept that there is reason and rationality in the human intellect, but I don't see how there can be reason and rationality in the world itself, independent of and separate from the human mind.

R: If God is indeed the Creator of this world, then God's rationality can be found in His creation, just as it can be found in the rationality of the human mind, which God created as well. Not only is the human mind consistent—because we were made by God in his image and likeness—but so is the world itself "consistent"—because the world came also from the hand of God.

You probably wonder: can science testify to it that there is some form of rationality in the world itself? Yes, science shows us again and again that the universe has an elegant, intelligible, and discoverable underlying mathematical and physical structure. The laws of nature are so orderly that they can be expressed quite precisely with mathematics. The idea that all matter is constructed from molecules and atoms that follow mathematical principles is quite surprising. This idea is also visible in scientific efforts that seek to "unify" scientific theories into more comprehensive structures. Such unification is only possible if there is a "consistent" structure behind the universe itself.

Unification is nothing new in physics. Think of the following cases in history. Newton's laws united earthly and heavenly bodies. Electromagnetism united electrical, magnetic, and optical phenomena. Quantum theory united particles and waves. Einstein's four-dimensional concept of space-time united space and time. And now a still unknown GUT might someday unite, unify, and combine Einstein's gravity and Planck's quantum theory. A mathematical unification of physical forces is reasonable, and it can only succeed if the universe seems to be susceptible to being unified. It proves again what the rationality of the human mind can do for us. This fact made even the astronomer Fred Hoyle, once an outspoken atheist, exclaim, "A common sense interpretation of the facts suggests that a superintellect has monkeyed with physics, as well as with chemistry and biology."

Perhaps this was even more astutely worded by the late astrophysicist Sir James Jeans when he said, "The universe begins to look more like a great thought than a great machine." Yes, Jeans spoke explicitly of a great "thought." This might be seen as a reference to the "great thought" of the Divine Intellect making for a consistent structure of the universe. And Stephen Hawking famously noted that all theoretical physics can do is give us a set of rules and equations that correctly describe the universe; however, it cannot tell us *why* there is any universe for those equations to describe. At least Hawking was pointing here to the larger enigma as to why there is something rather than nothing.

This shows us that even scientists uphold the conviction—consciously or subconsciously—that there is an intelligible plan behind this universe, a plan that is accessible to the human intellect through the natural light of reason and human exploration. It is the mind's rationality that gives us access to the laws of nature and the structure of this universe. Without rationality, we would not even know there are laws of nature. Thanks to the beauty of the human mind, with its intellect, we have the capacity to understand and explain the world the way it is. That's what I keep speaking of the beauty of a rational human mind.

The Beauty of the Soul's Voluntary Part

S: So far, you could explain to me rationality and its beauty. But when it comes to morality, you may have a much harder case to defend, since "morality" has become a rather ambiguous term nowadays. Some take that term as another word for ethics. But I don't think there is any beauty in ethics; it merely studies the various interpretations of morality. So, how can you still defend the beauty of morality?

R: I would define morality not as equivalent to ethics, but as our human faculty of making moral evaluations and decisions. Morality is about standards of bad versus good behavior in every single human being, whereas ethics is usually only about a code of ethical standards for specific groups of people (e.g., business ethics, medical ethics).

However, seen the way I understand morality here, there is a connection with the *voluntary* part of the human soul—parallel to the way rationality is connected with the intellectual part of the soul. The faculty of morality enables human beings to be moral beings, similar to the way the faculty of rationality enables human beings to be rational beings. But these two are not always in line with each other. The voluntary part of the soul is not always in agreement with the intellectual part of the soul. It is the will, says Blaise Pascal, which "dissuades the mind from considering those aspects it doesn't like to see." So, the voluntary part of the soul may sometimes overrule the intellectual part. This explains why some people are not *willing* to accept sound arguments coming from the mind. The will determines whether we do reject errors and deceptions, or rather decide not to.

What then is the main characteristic of morality? Morality is our main "tool" to get to what is "good," and not evil—just as rationality is our main "tool" to determine what is "true," and not false. Keep in mind, though, that morality is not another word for social behavior. Whereas social behavior can also be found in the animal world, morality cannot, as it adds a very specific dimension to social behavior—the dimension of what is good and what is evil. Social behavior in the animal world does not have that dimension. What then is it that makes morality so different from social behavior?

Morality is not about what the world *is* like, but about what the world *ought* to be like. It is not about how we *do* act, but about how we *ought* to act. It tells us what ought to be done—by us, as a moral *duty*, and towards us, as a moral *right*—otherwise a moral mistake would be made, similar to the way we can make rational mistakes. In brief, as moral beings we ought to do what we ought to do. C. S. Lewis put it this way, "human beings, all over the earth, have this curious idea that they ought to behave in a certain way, and cannot really get rid of it." From this follows that not everything that is thinkable or possible or reasonable—in rational terms or social terms—is also *permissible* in moral terms.

Apparently, moral behavior is not just a social phenomenon, but a distinctively moral phenomenon. The following simple example may help clarify the difference. Think about what is wrong with bullying (or any other kind of violent behavior, such as rape). As a biological feature, it may be very advantageous to have such a gene or allele. As a social strategy, it may be an effective strategy of banding together against someone else. But as a moral issue, it is plainly wrong. We have the *free will* to do what is right and refuse to do what is wrong. This gives morality an extremely specific and limited scope: righteousness. At the same time, it also has no bounds: it can be applied to an unlimited range of actions and situations.

S: I am surprised that you bring the human will into the discussion. Science is revealing more and more that what people call their "free will" is merely an illusion. What we think is a matter of free will is in fact determined by physical, chemical, and biological factors beyond our control. If so, that would be the end of morality as well.

R: I am not so sure that science is showing us what you think it does. It is rather a group of scientists who do so. Some geneticists, for instance—albeit a minority probably—are of the opinion that what some call their "free will" must, like anything else, be determined by neurons, and ultimately genes. In the words of the DNA co-discoverer Francis Crick, "your sense of personal identity and free will are, in fact, no more than the behavior of a vast assembly of nerve cells and their associated molecules." If that is true, then thinking that life is controlled by yourself would indeed be an illusion—merely under control of the stars, the laws of nature, your neural network, your DNA, or by anything else, but not yourself. This way of thinking is called *determinism*, yet another example of reductionism.

Scientists tend to agree with determinism—perhaps even understandably so. Science would not be possible if it did not assume that like causes have like effects and that the future depends on the past. The orderliness of this universe is a prerequisite for science, so we found out. If like causes do indeed always produce like effects, then the past appears to determine the future. Today's weather, for instance, depends on what happened yesterday, and so will tomorrow's weather depend on today. The French astronomer Pierre Simon Laplace probably worded this in all its consequences: "We may regard the present state of the universe as the effect of its past and as the cause of its future." In this view, we would all be in the iron grip of determinism—the future completely determined by the past.

This idea has been applied to many other areas of scientific research, including genetics. Even those kinds of behavior that we think are our own choosing—lifestyle choices, moral decisions, religious beliefs, and the like—made some scientists postulate a gene for whatever it is we decide, believe, or choose. No wonder then that new hypothetical genes just have been coming and going. Not surprisingly, we have been bombarded with new genes: a gene for homosexuality, a gene for schizophrenia, even a gene for religion—the list could go on and on. We could even invent a gene that makes one believe in the all-powerfulness of genetics—a gene for determinism, so to speak. That is where science borders on science fiction, or actually turns into it.

Although I must admit there must be determinism to a certain degree—like causes having like effects—I disagree with a rigid interpretation of determinism. The former acts like a technique, the latter as a worldview. The rigid interpretation ignores, for instance, the fact that the direction of something like a billiard ball on the pool table, or a golf ball on the golf course, is not only ruled by physical laws but also by human intentions—by players who have a certain goal in mind. In the midst of physical causes, the mind can create non-physical causes of its own. In other words, even in a world ruled by the law of cause-and-effect, there is also our own ability to be the cause of events all by ourselves.

That gives me reason to reject a rigid interpretation of determinism. Besides, believing in complete determinism takes us into a vicious circle: if there is only room for determined things in determinism, then the *claim* of determinism must be one of those determined things, too—so it is not something I can freely choose. Put differently, why debate free will if all the participants in the debate are already determined to either believe in it or not? Apparently, the claim of rigid determinism just destroys itself. Hence, there is no reason to defend it rigorously.

When we apply rigid determinism to the very issue of morality, as you suggested in your question, we are in for even more trouble. Doing so would give each one of us one giant alibi to dodge human responsibility. Once we declare ourselves no longer responsible for our moral decisions, we think we are off the moral hook. If all murderers had no choice but to commit their crime, it is unjust to blame them for anything they did. Some have come up with some quite creative ways of defending such alibis. All one needs to do is "geneticize" or "medicalize" bad moral behavior: The victimizer is no longer a person but rather a disease or pathology—a "disease" supposedly beyond moral control. A popular slogan says, "My hormones made me do this!" In this view, it is not hormones, or the like, that make us claim this, but rigid determinism does.

Therefore, I see no reason to go along with your idea that the notion of free will is just an illusion. And I have many people on my side. G. K. Chesterton, for instance, once said in a rather direct and simple way as is characteristic for him: if the world is fully determined, it makes no sense to say "thank you" to the waiter for bringing the mustard. To give thanks implies that something that did happen need not have happened. Even an eminent scientist such as the late Nobel-laureate and physicist Arthur Compton expressed his rejection of determinism very strongly when he said, "If the laws of physics ever should come to contradict my conviction that I can move my little finger at will, then all the laws of physics should be revised and reformulated."

Let me reword this with the terminology I used earlier: because "I-now" is always a pace ahead of "I-past," it is "I-now" that can take a next step, a step ahead of any physical determinism. "I-now" equals "I-as-a-subject," my "self," free to take a next step, a step ahead of any physical determinism, by acting as a new cause with a new effect. This means also there is still plenty of room left for morality. It is quite obvious that having a free will is essential for us to be masters of our actions—instead of victims—by allowing us to make our own choices and decisions in life, and even more so when it comes to morality. We were created free by nature, but not free to create our own moral universe. If God had not given us the gift of free will, we could not be rational and moral beings who can and ought to make rational and moral decisions. St. Thomas Aquinas worded this in his own characteristic way: "Free-will is the cause of its own movement because by his free-will man moves himself to act."

Humans are in control of those choices, at least to a certain degree. Human beings are free to choose whether to act or not to act—and in choosing, their choices cause them to act as much so as genes can cause them to act. Our destiny is not only in our genes or in other material factors, but also in our own hands—or ultimately in God's hands. We are free to choose between good and evil, but obviously our choices do not determine what is good.

S: Even if I agree with you that we do have a free will, I would still question whether morality itself is something real, or rather something imaginary. Our free will may not be an illusion, but our morality could still be.

R: True, moral issues are not settled by experiment; instead the answer to your question must be grounded in philosophy and metaphysics. Skeptical philosophers always argue that moral judgments are not judgments about the real world but merely expressions of our *feelings* about the world—which makes them surreal merely floating in the air. I would counter that this is just a view, an opinion they have, not a result of scientific investigation. In this view, moral laws are merely man-made laws. As a consequence, these skeptics argue that no one can impose one's personal feelings on others. How do they know this to be so?

A position like this is based on metaphysics. Metaphysics has to do with what is real and what is not. Even science cannot exist without some kind of metaphysics. It assumes, for instance, that causality is real, although it cannot scientifically prove it is. There is no way we can prove that like causes have like effects, but we presume that they do, otherwise science would collapse. That calls for a sound kind of metaphysics, based on common sense and on reasonable arguments. Chesterton said it right, "A society is in decay, final or transitional, when common sense really becomes uncommon." However, don't confuse "common sense" with "common science," for common sense comes before science can even begin. It tells us, for instance, that nothing can come from nothing, that like causes have like effects, and that the world around us is real, not an illusion.

In a similar vein, morality depends on metaphysics as well. Common sense tells us that there are also moral values and moral laws in life. People who deny this have no arguments to offer but merely enforce their own kind of metaphysics on everyone else, and by doing so they also create their own kind of morality. However, everyone knows, through common sense, that it is wrong to kill another human being. It is something we all have in common because it comes with human nature, with being human "in nature," regardless of ethnicity and race. When people lose their human nature, especially so in morality, we call them "inhuman."

As a matter of fact, ordinary people with "common sense" know right from wrong; that's one of the reasons why they can qualify as jurors in a court system. Even though some do kill other human beings, these people still know, deep down, that it is wrong to do so, but they were driven by some passion such as greed, revenge, or anger, which overtook their moral judgment. Passions pull human behavior down to its lowest level—it's like a psychological law of gravity similar to the physical law of gravity. Pascal was right, the will can persuade the mind not to consider those aspects it doesn't like to see.

When someone denies the existence of the laws of nature, such as the law of gravity, all one can do is invoke the principle of common sense. Common sense tells us there is some kind of *physical* order in nature. Gravity is not just a good idea—it's the law! Stones that fall today will also fall tomorrow. Although we cannot prove this today, we can still confidently assume it. We cannot even prove that all like causes have like effects. Similarly, when someone denies the existence of moral laws and moral values, all one can do is invoke the principle of common sense again. Common sense tells us there is some kind of *moral* order in life. For instance, if murder or slavery or any other kind of moral issue is wrong today, then it will also be wrong tomorrow.

Morality is ruled by moral principles, moral guidelines, moral laws, moral rules, moral commands, moral commandments, moral obligations, and moral values—call them whatever you prefer—but all of them together are often referred to as a "moral code," a set of standards prescribing righteous behavior. A moral code does not describe but it prescribes. Just as there is a physical order in this universe, so is there also a moral order. While the physical order is based on the laws of nature, the moral order is based on the laws of a moral code. The main difference is that physical laws *can't* be ignored, whereas moral laws *shouldn't* be ignored. Yet, in both cases, ignoring them has serious consequences.

Morality tells us what *ought* to be done—by us, as a duty, and towards us, as a right—otherwise a moral mistake is to be made. Therefore, morality is also about our rights and obligations, about which actions others owe us and which actions we owe to others. Duties and rights go hand in hand, so they have a natural reciprocity: the duty of self-preservation is also the right of self-preservation; the duty to seek the truth is also the right to seek it; the duty to work for justice is also the right to pursue it. In other words, "no duties" means "no rights," and vice-versa, "no rights" means "no duties." To mention a few examples: There is no duty to die, so there is no right to die; there is no duty to marry, so there is no right to marriage; there is no duty to have children, so there is no right to have children.

S: What then is it that makes morality so demanding that moral standards "ought" to be followed, as you put it? How can morality ever demand absolute authority? I would say, no other human being has the right to demand absolute obedience from me.

R: The moral laws of morality come with some very peculiar characteristics: they are unconditional, universal, timeless, absolute, and objective. Let me explain what that means:

1. Moral laws are unconditional. Most rules we are familiar with are conditional upon a certain goal—they are means-to-other-ends: if you want Y, you must do X; if you do not want to attain that goal, the rule does not apply. For instance, if you want to learn biking, you must do certain things; if you want to recover from a cold, you must do so-and-so, etcetera. Not so with moral rules and laws. They are unconditional: just do X, for you *ought* to do X—no matter what, whether you like it or not, whether you feel it or not,

whether others enforce it or not. Therefore, when it comes to morality, we cannot just pick whatever we want.

Morality tells us what *ought* to be done; it obliges us to pursue unconditionally what is good and right—no more ifs; no more questions asked. The legendary US President Abraham Lincoln put it well when he challenged the Nebraska bill of 1820 that would let residents vote on whether slavery should be legal in their state: "God did not place good and evil before man, telling him to make his choice." In brief, there is no "pro-choice" in morality. Put differently, choice does not determine what is right or wrong, but it is morality that determines which choices are right and which are wrong.

2. Moral laws are universal. This means they apply to *all* human beings—regardless of age, race, ethnicity, nationality, culture, or political affiliation. Morality is not regulated by popularity, interest groups, or majority votes, but it is universal in scope—it demands the same of everyone everywhere. Moral laws are as universal as the law of gravity is universal. As C. S. Lewis once said about moral values, "The human mind has no more power of inventing a new value than of imagining a new primary color."

Think, for instance, of the Nuremberg trials that took place after World War II—or of any other international court, for that matter. Seen from a purely legal point of view, it would not have been right, or even possible, to bring to trial and punish the Nazi perpetrators who had applied the civil laws that were created and implemented by a regime that had come to power through legal channels—for they were just "law-abiding" citizens following the law of the land. However, seen from a universal moral code perspective, their "lawful" actions were atrocities committed against humanity. To use a more specific example, we could ask about Nazi death camps: were they morally wrong, or were they just wrong in the eyes of nations other than Germany after World War II?

3. Moral laws are timeless. Those who object to the idea of a timeless morality might claim that moral rules and values have repeatedly been subject to change during the course of human history. However, this objection is based on a mix-up between moral *values* versus moral *evaluations.* Moral evaluations are our personal feelings or discernments regarding moral values and laws at a certain point in time. Moral relativists think that, in making moral evaluations, we create moral values in accordance with these evaluations. Hence, when evaluations change, the moral values and laws are said to change as well. However, moral evaluations are merely a reflection of the way we discern timeless moral values and react to them for the time being.

In other words, moral evaluations may change over time, but moral values do not—they are timeless and were always there, even when we did not discern them yet. The law of gravity, for instance, was already true before Isaac Newton discovered that law, but before him, we just did not know it yet. Something similar holds for moral laws and values. Slavery is a case in point. A few centuries ago, slavery was not evaluated as morally wrong, but nowadays it is by most people. The moral law did not change, but for a while many were "blind" for that law. Think of this: Although a visually impaired person cannot see the trees outside, the trees are still there, for the existence of the trees does not depend on whether the blind person perceives them or not. In a similar way, a morally blind person cannot see the moral laws and values out there, yet they are there; their existence does not depend on whether a person with moral blindness does perceive them.

4. Moral laws are absolute. Because of its absolute character, morality cannot be based on anything that is relative and non-moral by nature. Many, however, have tried to do just that. Utilitarianism, for instance, considers something morally right depending on its effects—that is, if it leads to "the greater happiness of a greater number of people." If this were true, we would not be morally obligated, but would only *feel* obligated. In this view, a moral code could be judged by measuring quantities and qualities in our direct surroundings. The problem of this view is that it treats moral laws as if they were *instrumental* laws—that is, relative laws which tell us what we need to do in order to attain something else. Once we consider as morally right anything that leads to more pleasure or more happiness, we are making

those acts instrumental to a certain goal or purpose. But moral laws cannot be treated that way.

In other words, there is nothing "useful" about moral laws. If anyone ever wonders why a certain act is good in a moral sense, we have no explanation to offer and cannot refer to other ends; all we can say is, "It's evidently right to act this way, and wrong to not act this way." It is a matter of self-evidence. without any exceptions. Killing a human being is always morally wrong; stealing is always morally wrong; lying is always morally wrong—no matter who you are and where you are, regardless of your status in society, and regardless of any personal circumstances. Moral absolutes are like unchanging rocks beneath the changing waves of feelings and practices. A morality of mere convention, of man-made and thus man-revisable rules of the social game, is not morality at all, but only a collection of customs and mores.

5. Moral laws are objective. The term "objective" here means that something is real and true, regardless of whether we know it to be true or not. In other words, moral laws are not invented by our minds, but they must be discovered in the reality around us. They are "out there" and exist outside of us, independent of our thoughts and wishes. They are not merely a matter of "different tastes," as some think they are. Yet, many nowadays tend to associate "values" in morality with the ever changing value of houses and stock, rather than treating them as something constant, absolute, and objective. Our secularized age has become crammed with "val-

ues"; for example, corporations and universities are proud to tout their "values." Besides, politicians often say certain policies go against their "values," but what they mean is that those policies go against their political wishes. As the Boston College philosopher Peter Kreeft astutely remarks, "God did not give Moses 'The Ten Values.'" In the words of the Catechism (1751), "Objective norms of morality express the rational order of good and evil." As a consequence, they are real—not just mental creations or a concoction of the mind.

Summarized, all five characteristics of moral laws—being unconditional, universal, timeless, absolute, and objective—are similar to the characteristics of the laws of nature on which the physical order of this world is based. They both are universal (applicable to everyone everywhere), absolute (without exceptions), timeless (even if we do not know the underlying law yet), and objective (a given, independent of us and of any human authority). They are objective, universal, timeless, and absolute standards—no matter whether we are talking rationality, in terms of true and false, or morality, in terms of right and wrong. Just as "truths are true," even when we do not know yet they are true, "rights are right," even though we may not realize yet they are morally right.

S: Even if I accept that moral laws can be all of these—unconditional, universal, timeless, absolute, & objective—

the question remains of how they can be that way. There must be an explanation for that.

R: I would answer that all the aforementioned characteristics of moral laws, rules, and values come from Heaven. Interestingly enough, even an atheist such as the late French philosopher Jean-Paul Sartre realized that there can be no absolute and objective standards of right and wrong, if there were no eternal Heaven that would make moral laws and values objective and universal. Being an atheist, though, he had to conclude that it is "extremely embarrassing that God does not exist, for there disappears with him all possibility of finding values in an intelligible heaven." As Sartre explained, "There can no longer be any good *a priori*, since there is no infinite and perfect consciousness to think it." Sartre realized very clearly that, by being an atheist, he also had to give up on morality. If there is no God, who is the source of all-goodness, then there cannot be evil either. Referring to God as the source of all good, St. Thomas Aquinas famously said, "Good can exist without evil, whereas evil cannot exist without good."

In his book *Co-Workers of the Truth*, Cardinal Joseph Ratzinger expressed a similar thought when he said: "if God does not exist, nothing will be as it is now; everything will proceed from emptiness and revert to emptiness. What we call justice will be mere caprice that we can rewrite as we will.... Whether or not God exists, the answer to that question will ultimately determine whether or not we are human beings, whether human dignity and true humanity and human justice can or cannot exist" The writer Fyodor Dostoyevsky summarized this well with a few words when he said that, without God, all things are permitted.

From this we derive that not everything that is thinkable or possible or reasonable is therefore also *permissible*—it may not be permissible from a moral perspective. Why is that? The reason is quite simple: it is God who made the standards of right and wrong unconditional, universal, timeless, absolute, and objective. When people debate what is right or wrong, they always appeal to an unstated higher moral standard. The beauty of the human mind, including its voluntary part, is that we have access to this eternal world of morality. Its beauty comes from its origin, made in the image and likeness of God. We ought to do what we ought to do—for heaven's sake!

S: Even if I accept what you say about morality, I would counter that what you call the faculty of morality didn't just arise from nowhere. As a scientist, I need to stress that this faculty must somehow have come along during the process of natural selection that humanity went through.

R: I think there are many reasons why the human faculty of morality cannot be the outcome of natural selection. First of all, how could natural selection ever make something *obligated*? How could a genetic code favored by natural selection ever make something mandatory? Natural selection cannot determine between morally good or bad results, only between success and failure. Natural selection is a process unfolding in the material world. However, there are no "oughts" or "rights" or "duties" in the material world. It is hard to see how non-moral causes such as genes and natural selection could ever produce a moral effect; they are of a completely different nature. A material process cannot make something obligated—perhaps effective or successful to some degree—but never obligated, let alone morally right or wrong. Access to a moral code, to a world of "oughts" and "rights" and "duties" seems to be a unique faculty which sets human beings apart from the rest of the animal world.

Also, natural selection "selects" genes that are successful or advantageous to their user. How could this ever explain that our moral code is based on a genetic code? I don't see how. If moral behavior were based on a genetic code, there would no longer be any need for a moral code as well—a moral code would be completely redundant. We would all act right by mere nature, forced by our genes, so there would no longer be talk of acting morally right or wrong. In contrast, I would argue that moral laws tell us to do what natural selection does *not* promote and what our genes do *not* make us do. Reality tells us that far too many people are willing to break a moral rule when they can get away with it. It is hard to believe that they are acting against their genes. Morality is about something that is outside the scope of biology—it is in fact beyond the reach of science.

Add to this that moral laws do not seem to have any survival value and therefore could hardly be the outcome of natural selection. Natural selection favors those who survive, but morality is not survivor-friendly. Moral laws such as "You shall not kill" or "You shall not commit adultery" could never make it through natural selection, for their offenders—the killers, the promiscuous, and the rapists—would be much more successful in reproduction than their victims or the ones who go by a moral code.

Apparently, morality and "survival of the fittest" do not go well together. Natural selection is about success at the *expense* of others; morality is about duties to the *benefit* of others. Natural selection *eliminates* the ones who cannot care for themselves; morality instead *cares* for those who cannot care for themselves. To put it more specifically, whereas natural selection is based on self-preservation at the cost of others, morality is often self-sacrifice for the good of others. When firefighters or soldiers on the battlefield die in the line of duty, they typically do not give their lives for self-preservation.

We must conclude from this that morality is *not* under the rule of natural selection, because the moral decisions we make are not "inborn" and not "useful," and therefore cannot be subject to natural selection. Hence, biology is blind to moral values, so it cannot possibly discern anything on its "blind spot." Morality can interrogate science, but science cannot question morality—it is beyond its reach. Therefore, science cannot monitor morality, as it is actually the other way around—morality should monitor science instead. In short, morality is not under the control of genes or natural selection, but under the control of the human will—another beautiful part of the human soul.

S: If you are right that morality does not come from our genes or from the history we went through as a species, then I need to ask you: where else then could the human faculty of morality come from?

R: It is a rather common conception that morality is merely a matter of human emotions. It is St. Thomas Aquinas in particular who counteracted that misconception. He based morality on reasons rather than emotions by mentioning the *natural law* as a foundation for morality. Its key idea is that moral laws are based on human nature, on the way human beings *are*—which is a metaphysical statement. In this view, morality is a function of human nature—the way we were created by God. That means *reason* can discover valid moral principles by looking at the nature of humanity and society, no matter where on earth. In other words, what we ought to *do* is not a matter of emotions but is related to what we *are*. Human beings are "by nature" social beings from whose nature flow their relations to family, fellow people, and society. They are not just joined together by artificial, contractual, and external ties. They share a common human nature despite differences in appearance.

From our common nature of being social beings follow all moral commandments., including the Ten Commandments. "You shall not kill," for instance, is based on the objective value of *all* human life and the need to preserve it. "You shall not commit adultery" is based on the real value of marriage and family, the value of mutual self-giving love, and children's need for trust and stability. We share these moral convictions to some degree with all of humanity. Every culture in history has had some version of the Ten Commandments. Yet, President Ronald Reagan was right when he joked, "I have wondered at times about what the Ten Commandments would have looked like if Moses had run them through the U.S. Congress..."

The claim of St. Thomas Aquinas is that the natural law is accessible to everyone endowed with the faculty of morality, which is every one of us. As a matter of fact, there is not a great deal of difference between a Christian morality, Jewish morality, Hindu morality, Muslim morality, Buddhist morality—although there is a great difference in these religions. C. S. Lewis, for example, published a list of universal moral principles which he called "Illustrations of the Tao or Natural Law." As Peter Kreeft puts it, "We find similar morals, beneath different mores." In other words, what the Catholic

Church has to offer is not some outlandish faith based on emotions, but a faith that comes with reason. Most of all, she has the natural law to offer, which has always focused on the common good of society and is ingrained in human nature, because we are social beings by nature.

This explains also why a non-Catholic such as Martin Luther King Jr. could refer to the natural law when he said, "A just law is a man-made code that squares with the moral law, or the law of God. An unjust law is a code that is out of harmony with the moral law. To put it in the terms of Thomas Aquinas: An unjust law is a human law that is not rooted in eternal law and natural law." In other words, the natural law is an eternal law that our souls have access to thanks to our beautiful minds and wills.

S: I often hear people say, "Always follow your conscience," or "Never disobey your own conscience." I would say that idea looks like a better rule to me than using your conception of morality in terms of the natural law.

R: It may seem that way, on condition that we know how to understand the term "conscience." Our conscience has often been compared with technical devices we are all familiar with—such as a compass, a global positioning system (GPS), a barometer, an alarm, a gas gauge in a car, and the list goes on and on. These analogies are right about one thing: our conscience is indeed some kind of monitoring device—it monitors what is good or bad, right or wrong. But what such analogies mask is that these devices are merely tools that may not always work properly or may even fail entirely—and the same may be said about our conscience.

A real compass, for instance, functions as a pointer to the magnetic north because the magnetized needle typically aligns itself with the lines of the earth's magnetic field—that is, with something outside itself. But there are some caveats: it should not be used in proximity to ferrous metal objects or electromagnetic fields, as these can affect their accuracy. At sea, for example, a ship's compass must be corrected for errors, called deviation, caused by iron and steel in its structure and equipment. Something similar holds for a gauge. The gas gauge in your car may no longer go down because it is broken, yet the tank may be almost empty. And your GPS system may not work when something obstructs the connection with the satellite high above your head.

In other words, a person's conscience may indeed function in ways analogous to a compass or GPS, but these "monitoring" tools must be monitored themselves and be aligned to an outside source—as does our conscience. Just as a compass must be "aligned" with the earth's magnetic field and protected from surrounding interference, and a GPS system must be "aligned" to the right feed from satellites high in the sky, so too needs a human conscience constant alignment to something like the natural law.

But also keep in mind that, because our moral compass can sometimes fail, we cannot conclude there is no right direction at all. Just like with math, we can get our sums wrong, but just because we may have different calculations, it does not follow that there is no right answer at all. So, the question then is: what is the "right math" in morality? What is the right feed for our conscience? How do we properly align it? What do we align it to? In short, how do we calibrate our conscience? Only when we know those answers can we say, "Never disobey your conscience."

The Catholic Church would say human beings were created with a moral compass pointing to, not the magnetic North, but to the "Above"—to a place where justice reigns and moral laws reside. Therefore, our conscience is not a private "compass" that determines its own North Pole. It must be aligned to the one and only real "North Pole Above"—otherwise we can easily go off track. So obviously, in response to your question, there is more to it than having a conscience and following one's conscience. When people say, "Never disobey your own conscience," they forget one can do things "in good conscience," but also "with a bad conscience." Apparently, a conscience on its own can be good as well as bad. That's why one can do things "in good conscience," but also "with a bad conscience." A person's conscience can in fact be so darkened that the Psalm can say about a sinner, "He so flatters himself in his mind that he knows not his guilt" (36:3).

In summary, morality is the hallmark of humanity and sets us apart from the animal world. It comes with human beings who were made in God's image and likeness, so that the human mind and will are a take-off of the Divine Mind and Will. If God is all-good and all-righteous, then we ought to be good and righteous too. We were made in such a way that we are able to do good and ought to do good. Isn't that beautiful? Well, yes, that is the beautiful faculty of morality given to us.

What Has Happened to the Beauty of the Soul?

S: If the human mind and will were really a take-off of the Divine Mind and Will, then all of us should have a perfect mind and will. Isn't that what follows from what you are suggesting? However, you don't have to be a rocket scientist to see many flaws in the so-called beauty of the human soul.

R: Yes and no. Yes, I do defend that we have beautiful *minds* to help us understand and beautiful *wills* to help us choose. Yet, something has gone awfully wrong with the human mind and the human will, which has made them a flawed reflection of what God intended them to be. Hence, we are no longer unfailing in using rationality and morality. Often, we do not understand correctly, nor do we choose rightly. How come?

Whereas the creation myths of many cultures hold that good and evil are inherent elements of human nature and the world order—by just assuming that is the way things were made—the Book of Genesis corrects this erroneous view by revealing that evil is not rooted in creation, but in humanity's abusive decision to turn away from God, from one another, and from God's created order through sin. Because of the so-called *Fall*, the image of God has been corrupted, and Adam and Eve's sin has been passed on to the whole human race, which is usually referred to as "Original Sin." Although the goodness of their original creation remains standing, humans are now flawed images of God. So, the imperfection and fallibility of the human mind and the human will are not God's doing, but our own. When you ask me, what is wrong with human beings, my answer is rather simple, but direct: they are descendants of Adam and Eve who passed on to them the Original Sin.

I think the famous Catholic philosopher and physicist Blaise Pascal (1623-1662) worded it well in his *Thoughts* (*Pensées*): "It is dangerous to show man in how many respects he resembles the lower animals, without pointing out his grandeur. It is also dangerous to direct his attention to his grandeur, without keeping him aware of his degradation." Pascal hit the nail right on the head. Human beings have received grandeur from God as well as degradation from themselves. The Book of Genesis reveals to us that something did go wrong with humanity. But be aware: it is the Fall, not God, that caused our degradation. Because we lost our original rationality and morality, we are easily able to err in our rational and moral judgments and decisions.

St. Augustine tells us in his *Confessions* the famous story of how he stole pears from an orchard as a child. He astutely remarked about his stealing, "I had no wish to enjoy what I tried to get by theft; all my enjoyment was in the theft itself and in the sin." This is an example of the perverse will that chooses evil for its own sake—a consequence of Original Sin in which the will rebels against the intellect and becomes a slave at the command of passions. Just as the good can be loved for its own sake, as something intrinsically desired, so evil can be willed for its own sake, too. That's when someone's moral sense—and conscience, I might add—becomes desensitized, making evil appear to be good.

Do not just stare at the odd example St. Augustine uses. It shows what happens all the time. Some lie frequently just for the mere enjoyment of the lie itself. Some use guns just for the mere enjoyment of killing in itself. Some bully others just out of pure enjoyment. Some ridicule the religious beliefs of others just for the pleasure it gives them. And the list goes on and on. It is the sin itself that gives them enjoyment.

Reality tells us that far too many people are willing to break a moral rule when they can get away with it. Too many parents ignore what some think is an "inborn" responsibility of parenting. Too many spouses violate the sixth commandment, "You shall not commit adultery." Too many folks also violate the fifth commandment, "You shall not kill." When it comes to moral laws, everyone knows about them and yet everyone breaks them repeatedly. For instance, mothers who abandon their newborns are not so unusual, because maternal responsibility is a moral law that can be ignored. Obviously, morality can be overshadowed by the tyranny of unruly passions as well as by destructive doctrines. That is a reality because of the Original Sin. People who live in sin tend to be guided by a darkened mind and a corrupted will. They are prone to making errors in their rational and moral judgments and decisions.

S: How can you defend a concept as murky as "Original Sin"? Sorry, the scientist in me tells me this idea is not only unscientific, but it even goes *against* science. How can something be hereditary without being genetic?

R: I would say, you don't have to be a rock scientist or a brain surgeon to know there is something fundamentally wrong with humanity, something that is "rotten" to the core. There is probably no better term to capture this than the term "Original Sin"—it is not genetic, yet it is hereditary, passed on from generation to generation. C. S. Lewis went as far as arguing that the existence of Original Sin is perhaps one of the most obvious facts of human life, even so for nonbelievers. And G. K. Chesterton called it "the only part of Christian theology which can really be proved." What the

Bible emphatically adds to it is that this is not God's doing, but ours—and it goes back to the very beginning of humanity and has been around ever since.

The Bible tells us that it all began in the minds of Adam and Eve, who refused to be "under God," but instead deemed themselves to be "next to God," or even "above God." If Eve and Adam had known nothing of good or evil before they had taken of that tree, how then could they be held responsible and be punished? Here is how C. S. Lewis describes it, "They wanted some corner of the universe of which they could say to God, 'This is our business, not yours.' But there is no such corner." Guided by the lies and deceptions of Satan, they thought some corner of the universe should be entirely theirs. By giving in to the divisive influence of Satan, their minds and wills became divisive too. The divide separating God and man had suddenly become widened. That's how the brokenness of human nature came along. It is this moment in early human history that started the Original Sin—something inherited without being genetic. How can that be? Well, think of this: if you come from a dysfunctional family, chances are you will start another dysfunctional family. You "inherited" a state of being "dysfunctional," so to speak.

Humanity—starting with the first human beings—wanted to be like God by making its own standards of rationality and its own standards of morality. The Bible uses the following words for this event: "But the serpent said to the woman, 'You will not die. For God knows that when you eat of it your eyes will be opened, and you will be like God, knowing *good and evil.*' So when the woman saw … that the tree was to be desired to make one wise, she took of its fruit and ate; and she also gave some to her husband, and he ate" (Gen. 3:4-6; italics are added). After them "eating from the tree," we all became accomplices.

Why is the doctrine of Original Sin so pivotal in Catholic Faith? If there is no Original Sin, then there is no need for redemption; if there is no need for redemption, then the Cross is a hoax; and if there is no Cross, then the whole economy of Salvation through the Incarnation is up for grabs. The Catechism (389) puts it this way, "The Church, which has the mind of Christ, knows very well that we cannot tamper with the revelation of original sin without undermining the mystery of Christ." Simply put, we do not need to be saved from a ship that is not sinking. Yet, we are sinking!

With the coming of God's Son, who alone can remove Original Sin, it will be possible for a redeemed humanity to live eventually again in an ideal world free from the corruption of soul, mind, and will—which is the Kingdom of Heaven, a broken world restored. Only the death and resurrection of Jesus can restore our original image in the likeness of God. It all revolves around Original Sin, the endemic state of a broken world. The God-Man Jesus went to the full depth of suffering and sin on Golgotha by identifying himself with our suffering in order to eradicate the effects of sin. This is the mystery of the Incarnation, and thus of the Crucifixion—the fact that the God-Man Jesus comes to us, weeps with us, and suffers with us, and thus redeems us. God became man to redeem us, not from our mistakes and weaknesses but from our *sins*.

S: How then, in your view and the Catholic view, has the Fall corrupted and darkened the mind of human beings?

R: We were created with a sound mind, allowing us to discover what is true and what is false. However, the Fall has darkened and corrupted the human mind in its search for truth. Truth is a key issue here. We were created with direct access to the truth, but since the Fall this direct access has been obstructed by ideologies that block our access to the truth.

We mentioned already several of those ideologies, more specifically in connection with science. Materialism makes us believe that truth can only be found in material things. Reductionism wants us to believe there is only truth in nothing-but-ery. Scientism tells us that science has the only tool to find the truth. What these ideologies have in common is that they do not accept any other truths. They are ideologies that blind us to see the real, full truth. C. S. Lewis once appropriately spoke of "willful blindness." I argued already that they are not products of a sound mind, but of a darkened and warped mind.

But the human mind has also been obstructed by ideologies with a much larger scope. I want to mention only the two arguably most devastating ones: skepticism and relativism, both of which came forth from darkened minds as well. Let us start with the first one, skepticism. In its general form, it questions all truth claims anyone might have, so it basically claims that there is no direct access to the truth, perhaps even no access at all. Skeptics are "masters of suspicion." Granted we often need to eliminate errors to get to the truth, still our ultimate goal is not to avoid errors, but to gain truth—to know rather than to know what we do *not* know. Skeptics, on the other hand, make it their final goal to avoid errors, in denial of the fact that eliminating errors is only a means to getting to the real truth. Skepticism is at best a method to avoid errors, but it never leads to truth—it's an activity, not an achievement. It uses a silly slogan like "It is better to travel hopefully than to arrive"—which, as C. S. Lewis pointed out, is nonsense, because if you believe this, then you never travel hopefully because there is no hope of arriving. It is like the horizon: at best approached but never reached.

No wonder, the commonsensical idea of "common sense," for instance, is typically considered illegal in skepticism; there is no common sense in skepticism. Yet, even those who swear by the "trial and error" method must admit that errors only exist by the grace of truth. The philosopher Gilbert Ryle astutely remarked that there can be no counterfeit coins without genuine currency. Well, truth is the "genuine currency" revealed by the beauty of the human mind.

S: In addition to skepticism, you also mentioned relativism as a product of a darkened mind. I wonder what it is then that makes relativism a corrupted ideology.

R: Relativism is certainly not a recent invention. The ancient Greek historian Herodotus said in his book *The Histories*, "Everyone without exception believes his own native customs, and the religion he was brought up in, to be the best." Then there is another ancient Greek, Protagoras, who is known for the expression "Man is the measure of all things"—which must necessarily include any truths a person may claim. Pontius Pilate would follow soon with his legendary question, "What is truth?" Doesn't that sound similar to what Adam and Eve thought: "your eyes will be opened, and you will be like gods"? Although the idea of relativism may have a long history, it was never popular or widespread, but nowadays it is receiving more and more traction.

People who swear by relativism—let me call them "dedicated relativists"—claim that there is no absolute truth—there are only *opinions* about what is true, depending on one's point of view. Relativism denies, or perhaps just doesn't want to acknowledge, that there is something like absolute, objective truth—truth that cannot be denied or declined by anyone. Relativists claim there is no such absolute truth, because they always think about truth as something relative to a particular frame of reference, such as a particular language or culture or tradition or religion. Whatever is true for me may be false for you; whatever is right for me may be wrong for someonc else. It is basically the idea of "anything goes." We have only mere opinions about what is true and false in rationality, or right and wrong in morality. Taken to its extreme: gravity may be true for you but not for me; killing may be wrong for you but not for me. It looks like a very democratic idea.

What is wrong with relativism? Relativists swap truths with opinions. They hail "freedom" after reinterpreting it as, what the philosopher D. C. Schindler sneeringly calls, "freedom from reality." Truth is no longer a matter of facts but of opinions, convictions, and feelings, detached from reality. Does that changeover matter? Well, if the world is in fact not flat, then someone's opposing opinion will not make it flat. Opinions do not trump truths. As we said earlier, truth is truth, even if you do not accept it; and untruth is untruth, even if you claim it. Truth is not something we create or invent on our own, but something that we try to "capture" in the real world. Truth is the conformity of mind and reality—a match between thoughts and reality. Therefore, truth is truth—for everyone, anywhere, at any time.

If relativism were true, then there could no longer be any dispute about what we claim to know, for we would in fact not know anything. The French philosopher Jacques Maritain said it right, "The sole philosophy open to those who doubt the possibility of truth is absolute silence—even mental." That also marks the end of rationality. Peter Kreeft makes a powerful conclusion, "Relativism is not rational, it is rationalization. It is not the conclusion of a rational argument. It is the rationalization of a prior action." I would say it is in fact the outcome of a warped mind—another result of the Fall.

But there is more wrong with relativism. The problem with a claim like "There is no such a thing as objective truth" is that it claims to be objectively true. We cannot make such a sweeping assertion without also asserting, implicitly, that there is such a thing as objective truth after all. So, relativists end up with a paradox: if their claim of relativism is true, then it becomes false. This issue has become known from the famous Liar Paradox: Epimenides, a Cretan, says "All Cretans are liars." Is Epimenides telling the truth or is he not? If he is, he is not; if he is not, he is. Whatever answer we choose, we are led into contradiction. Is this detrimental to the doctrine of relativism? I would say it certainly is. If nothing is true, how then can the sum total of nothings be sublime? Yet, relativism has invaded, as a Trojan horse, the mind of many nowadays, blinding them for the truth. That's what has happened to the beauty of the human mind.

S: I think there is still one area of human knowledge that relativism has not been able to invade: science. Science may be the last bullwork, where truth is still honored as a key issue. If you acknowledge that, aren't you in fact admitting that science—and only science—must eventually be is on its way to the final truth?

R: No, not at all! With your suggestion, you are in fact proclaiming another misguided ideology, which we called scientism before. What I claim in response to this is that there are facts and truths other than scientific facts and truths. Thinking that only science "knows," and that nothing or no one else does, is not a scientific fact but rather a belief, an ideology. Once we reject this idea—and with good reason—we must accept that there may be other truths than scientific truths.

However, they may be of a different nature. They differ, for instance, in the way we get to those truths. As a result, science and religion convey two different kinds of truth—let's call them "natural truths" versus "supernatural truths," or "earthly truths" versus "heavenly truths." Natural truths are backed by human authority, supernatural truths are backed by Divine Authority. Natural truths we do believe because science tells us they are true. Supernatural truths, on the other hand, we believe because God tells us they are true. Another way of distinguishing these two kinds of truth is this: There are "natural" truths that come to us by experience or experiment versus "supernatural" truths that come to us through Revelation and Scripture; or also, "human" truths that we gain on our own versus "divine" truths that we receive from Heaven; or also, truths we can see through our physical eyes versus truths we discern through the eyes of faith. Ultimately, of course, all truths come from God. That's obvious for supernatural truths—they come directly from "above." But even the natural truths come from God. Everything in nature has been created by God and could not exist without Him.

Take, for example, these two truths: human mortality is a natural, earthly truth, whereas human immortality is a supernatural, heavenly truth. Or think of this one: the Crucifixion of Jesus is a natural, earthly truth, whereas the Resurrection of Jesus is a supernatural, heavenly truth. Another example of a natural truth would be heliocentrism—in the sense of the Sun being the (gravitational) center of the Solar System—whereas an example of a supernatural truth would be Christocentrism—in the sense of Christ being the (spiritual) center of the universe. Natural truths we can often claim "in the name of science and with the authority of science"; supernatural truths, on the other hand, we claim "in the name of Jesus and with the authority of Jesus."

The distinction we made between two different kinds of truths should not obliterate the fact that they both deal with truths and facts, no matter whether they are discovered or revealed. Science and religion are, in the words of Francis

Collins, "like the two unshakable pillars, holding up a building called Truth." What both kinds of truths have in common is that truth is truth, even if you do not accept it; and untruth is untruth, even if you claim it. St. Augustine famously said, "All truth is God's truth." As much as it is a matter of fact that the earth is not flat—believing that it is flat does not make it flat—so it is also a matter of fact that God exists—believing that he doesn't exist, does not make him disappear. Truth is truth, although we may not fully understand or capture the truth yet. What unites both kinds of truths—natural versus supernatural—is that they both need to go through the test of reason and cannot violate rationality. To think that only science knows the truth is another example of a warped, misguided mind. Yet, secular historians, secular psychologists, and secular scientists all band together to dismiss the power of the supernatural.

Claiming that religion has its own source of knowing the truths does not mean, though, that all religions are of the same quality. Yet, many nowadays think they are. What these people basically defend is some form of relativism in matters of religion—thus claiming that all religions are worth the same when seen within their own cultural setting. Ironically, however, relativism makes each religion absolute—its own authority in matters of truth. This is like saying that what is "true" in chess is also "true" in checkers. The fact remains that relativism is basically flawed and has darkened the beauty of the human mind and its rationality. We could have known better, but do not always do so because of the Original Sin, which has darkened our minds.

S: You told me before that something has gone wrong with the human soul. So far, you have mostly done so for the intellectual part of the soul, our sense of rationality. But what would you say about our sense of morality? How has that been infected by false doctrines?

R: Just as rationality, which is the intellectual part of the human soul, can get twisted, when the mind goes on the wrong track, so can morality, which is the voluntary part of the human soul, get warped, too, because the will can be corrupted each time the will follows the wrong track. Both mind and will have the potential to get distorted and corrupted. Just as the intellect of many has lost the discipline of rationality, so has the will of many lost the discipline of morality. As a matter of fact, false doctrines not only corrupt the way we think in rationality, but also the way we act in morality. That is why the Psalm warns us for the corrupt stubbornness of the human will when it says, "Harden not your hearts" (95:8). Let me explain this in more detail.

How has our sense of morality been tainted with false ideologies? I mentioned already the ideology of Gnosticism which makes it look like souls are trapped in dungeon-like bodies and therefore want to be released from their prison. Another example is the make-believe doctrine of genderism, mentioned earlier, which infects people with a darkened mind and a corrupted will as to who they are and how to act. But perhaps the best example of a misguided ideology is the doctrine of relativism again, which may even be at the root of the other misconceptions, not only so in rationality but also in morality. In all such cases, we have thrown off God's reality—about life, sex, gender, etc.—and have tried to construct our own, just like Adam and Eve did in Paradise.

Relativism not only declares us unable to know what is true and what is false, but it also declares us unable to identify what is right and what is wrong—in denial of the fact that there are moral truths we should fight for as well as rights we are willing to die for. In the domain of morality, relativism claims that morals are no longer considered to be based on unconditional, universal, timeless, absolute, and objective standards of right and wrong, but on personal opinions and preferences. G. B. Shaw, for instance, became a spokesman for this, when he declared them just a matter of "different tastes," of what the Washington Post columnist George F. Will once sneeringly described as "the moral equality of appetites." However, those who defend this view are imposing their own philosophy of morality on everyone else. They create their own moral world and put a veto on any view dissenting from theirs. They make morality a matter of personal emotions and values in exchange for the moral values and rules which God has ordained.

In this erroneous view, we are no longer under any moral obligation—because there simply is no such thing—but we only *think* we are. The objectivity of morality has become a *sense* of objectivity, which is in essence a matter of subjectivity. We are not really obligated to act morally, but we may only *feel* we are. The destructive doctrine of relativism substitutes the universal moral code based on the natural law and on the makeup of human nature with a multitude of private moral codes based on highly personal choices, interests, and beliefs. We are supposed to be creatures whose nature it is to have *no* nature. If no one believes in anything moral, then all of us are free to live our own lives and do our own thing just the way we like.

The most amazing part of relativism is that it can never be consistent. It is ironic to see how those who deny that morality has any absolute authority still hold on to at least one moral absolute that says, "Never disobey your own conscience." So, moral relativists should then ask themselves the question as to where the absolute authority of a human conscience comes from. The answer is rather simple: this authority comes from themselves. It is the age-old story again: "your eyes will be opened and you will be like gods, who know good and evil" (Gen. 3:5). Relativists proclaim there is no moral authority other than their own based on their personal preferences and beliefs. Apparently, everyone has something they *know* is morally wrong, even relativists do. Yet, there is no way we can validly justify that our act was morally right by claiming our conscience told us so—even Nazis could make that claim to justify their atrocities. I think we discussed that point enough.

Besides, another ironic part of relativists is that they are very selective in their moral choices and preferences. They use relativism to be very judgmental about things they do not like, such as racism and slavery, whereas they consider abortion okay. But their moral preferences have nothing to be based on. Had the slaveholders won the Civil War, then we might today view slavery as an admirable institution. Or applied to more recent times: had we lived in Nazi Germany, we would all be replicas of Edith Stein or Maximillian Kolbe; we would never have been infected by the nationalist and socialist disease. That is in essence what moral relativism makes us believe.

One could even argue that the famous Golden Rule can be seen as another example of relativism. In its negative form, it says, "Do not do unto others as you would not have them do unto you." Its positive form is "Do unto others as you would have them do to you." Obviously, this rule is not based on anything universal, objective, and absolute—although it is also mentioned in the Bible (Tobit 4:15; Matthew 7:12; Luke 6:31). Instead, it refers to what you wish to be done or not to be done to you. So, it is in essence very self-centered and lacks any further foundation.

Some relativists have gone as far as replacing the moral code of the Ten Commandments with what sociologist Alan

Wolfe calls a new, eleventh commandment, "Thou shalt not judge." Isn't it amazing how even moral relativists come up with moral statements that everyone is supposed to accept as universal and absolute? It is an entirely inconsistent move on their part. If we say that everyone must be tolerant to everyone, except to people who are intolerant themselves, then we are enforcing the intolerance of tolerance. If we say that everyone must be inclusive to everyone, except to those who are not inclusive themselves, then we are enforcing an exclusion from inclusion. In all such cases, we may end up with a demand or rule that is quite dictatorial and authoritarian, allowing no room for anything but itself. It grants to some the authority to use their "veto power." Ironically enough, the only thing all relativists do avoid at all costs is the dreaded "N-word"—the Natural Law with its unconditional, universal, timeless, absolute, and objective standards of right and wrong.

It should not come as a surprise then that Pope Benedict XVI spoke of the "Dictatorship of Relativism" in our modern society. In April 2005, in his homily during Mass prior to the conclave which would elect him as Pope, then-Cardinal Joseph Ratzinger said, "relativism, which is letting oneself be tossed and 'swept along by every wind of teaching,' looks like the only attitude acceptable to today's standards. We are moving towards a dictatorship of relativism which does not recognize anything as certain and which has as its highest goal one's own ego and one's own desires."

S: Your arguments against the role played by relativism in destructing morality do sound reasonable. However, I cannot believe that there is nothing more to say about this. I see also the strong impact that new scientific developments have had on the morality discussion. I am referring particularly to a branch of biology that has become known as sociobiology.

R: You are right. But we do have a problem here. We talked earlier about self-awareness, self-transcendence, self-identity, and other important features of the human "self." It is in fact this very philosophical concept of "self" that science has hijacked for its own purposes and is trying to capitalize on, especially so in sociobiology.

Sociobiologists have used the process of natural selection to show that all organisms try to promote themselves—that is, their own "selves"—in the struggle for survival. Since natural selection is considered a matter of balancing the benefits against the costs, survival of the fittest would indeed be a way of promoting the "self" of the fittest at the cost of the "selves" of others. Some sociobiologists even extend this idea to genes, claiming that it is ultimately the genes that show this "selfish" behavior—another case of reductionism. They famously mention the example of sterile worker bees in a colony "unselfishly" helping the queen raise her own progeny. They would explain this behavior as a form of helping one's close relatives, because these carry DNA in their genes that is very similar to one's own DNA, so they refer to these genes as "selfish genes." Since natural selection is a matter of balancing the benefits against the costs, this would indeed be a way of promoting one's "own" DNA by diminishing one's own offspring (a cost) while increasing the offspring of one's relatives (a benefit).

Thoughts like these enticed the British biologist Richard Dawkins to write his book *The Selfish Gene*. What is going on here? First of all, speaking of "selfish genes" is as silly as speaking of "selfish atoms." It is merely a rhetorical trick used by Dawkins to make his case. Second, Dawkins shrewdly manipulates the terms he uses for his explanations. He borrows the term "selfishness"—a term that makes perfect sense in a philosophical and moral context—then applies it to genes in genetics, and then he uses the "selfishness of genes" to eliminate any kind of moral behavior as being a mere battle of selfish genes. What we have here is another case of nothing-but-ery. The "self" that biologists like Dawkins refer to has nothing to do with the "self" we use when speaking of self-awareness, self-transcendence, self-identity, and other important features of the human "self."

In contrast to these claims, I would stress that the biological metaphor of "selfishness" is only possible because we already have an understanding of human behavior in a moral and philosophical context. Dawkins turns things upside-down: he treats the biological metaphor of selfishness as the real thing, and then uses it in turn to discard human moral behavior as being unreal. He makes the gene the basic unit of "selfishness." However, the language of selfishness does not make sense in genetics, because genes cannot be selfish or unselfish the way human beings can be selfish or unselfish. Yet, he uses the expression "selfishness of genes" to get rid of any moral behavior in the world of human beings.

The assessment of the anthropologist Donald Symons might be right on target: "[T]he rhetoric of *The Selfish Gene* exactly reverses the real situation: through [the use of] metaphor genes are endowed with properties only sentient beings can possess, such as selfishness, while sentient beings are stripped of these properties and called machines.... The anthropomorphism of genes... obscures the deepest mystery in the life sciences: the origin and nature of mind." Hence, we are back to where we started—the beauty of the human mind and human will that comes with a human soul.

S: Yet I maintain that science must have something to do with morality. Or are you trying to tell me that science cannot explain morality at all?

R: Yes, I do think that science cannot explain morality. As I said earlier, morality is not about what is physically possible, but instead it is about what is morally permissible. The term "permissible" does not and cannot exist in the vocabulary of science. Take this comparison: it is not up to science to determine what is true or false in mathematics, but neither is it up to science to determine what is good or evil in morality.

Those who deny this are actually declaring that everything that is physically possible is therefore automatically also morally permissible. In that erroneous view, science determines morality. But that would be detrimental to morality, making it a servant of science, making whatever is possible automatically permissible—no further questions asked. However, morality cannot be at the mercy of science, because morality has its own unconditional, universal, timeless, absolute, and objective standards of right and wrong—all of which combined make for a set of standards prescribing righteous behavior. The situation should in fact be reversed: science should be under control of morality, not vice-versa.

As I said earlier, morality can interrogate science, but science cannot question morality—it is beyond its reach. Therefore, science cannot monitor morality, but it is rather the other way around—morality ought to control science instead. Nazi-doctors such as Joseph Mengele show us what happens when morality does not control their scientific research. This means there are two rather different ways of looking at "good science." Science can be "good" judged by methodological criteria—which are good-for-the-purpose—but it can also be "good" according to moral criteria—which are good-as-such. In the case of methodological criteria, we are dealing with relative rules or norms; they indicate whether a certain kind of science is good in terms of current methodological criteria. "Good" science in this sense is "methodologically sound" science meticulously following methodological guidelines. However, there also exists "good" science in another sense—in relation to certain moral rules and values, that is. Science practiced in Nazi camps was not "good" science in moral terms, although it may have been methodologically sound. That is where a moral assessment is needed.

We must conclude from this that scientists can no longer hide themselves in their "ivory tower," free from any moral assessment. Albert Einstein made an important distinction then he said you can speak of "the moral foundations of science, but you cannot turn around and speak of the scientific foundations of morality.... [E]very attempt to reduce ethics to scientific formulae must fail." Morality can interrogate science, but science cannot question morality—it is beyond its reach.

S: You explained to me so far where the beauty of rationality and morality comes from. But you did not explain where the devastating doctrines that darken rationality and morality could have come from, other than from Adam and Eve. How could the eating of some fruit cause the rise of such devastating doctrines? That fruit must have had some psychedelic chemical acquired upon consumption.

R: You are right, it is hard to believe that the Original Sin was caused by the Fall of two people who ate from one particular tree in Paradise. Many questions abound. Couldn't they have controlled their appetite more carefully? Are they the only ones to be blamed for all that happened after? That seems to be a rather primitive, outdated explanation of why there is so much evil in the world. However, there is much more the Bible wants to tell us. This Biblical explanation has a much deeper layer. It gives us a look behind the scenes of what is going on in this world.

It is not just about these two people, but about a much more powerful force behind them. This force is described with words such as serpent, snake, or dragon. But do not be deceived by such terms. They all ultimately represent evil and are identified as Satan in the Book of Revelation (Rev. 12:9): "The huge dragon, the ancient serpent, who is called the Devil and Satan, who deceived the whole world, was thrown down to earth, and its angels were thrown down with it." What is playing out in the deceivingly called primitive description of the paradise story is a battle going on between God and Satan, between the source of all that is good and the source of all that is evil. The source of all that is good is opposed and discredited by the serpent, the actual source of all that is evil.

The confrontation between good moral doctrines and evil moral doctrines in history is a repetition of the conflict playing out in paradise between the voice of God and the voice of the serpent, Satan. This confrontation in Paradise is just the beginning of a perpetual warfare of the forces behind what is true and good versus the forces behind what is false and evil. Do not forget how it took two giants in more recent history, motivated by their religion, to defeat the evil empire of the communist Soviet Union: Pope John Paul II and Ronald Reagan.

All of us are warriors in this warfare, taking sides with either the good forces or the bad forces. Satan plays an important role in Christianity, not only through Adam and Eve, but also in people like Judas Iscariot. St. John says about Judas, "the devil had already put it into the heart of Judas Iscariot, Simon's son, to betray him" (John 13:2). This does not take away from Judas any responsibility, let alone his human freedom. It is everyone's personal decision to let either God or Satan guide and steer one's actions. All people can sell their souls to either one—that's up to them to decide on an individual basis.

As a matter of fact, all destructive doctrines we talked about—materialism, dualism, reductionism, relativism, and skepticism, to name just a few—could very well be part of a much larger picture—a cosmic warfare, so to speak, between Good and Evil, between God and Satan. It is God's aim for each one of us to attain Heaven after death, whereas Satan's aim is to ensure that as many people as possible miss that eternal goal. Do not take God and Satan as two eternal principles locked in permanent conflict as they are in Gnosticism, for Satan and other demons are fallen Angels who were originally created good, but they decided in their freedom to go against their Creator and become evil. The Bible makes it very clear that Satan is a reality; and evil is something real to watch out for. God can do so much good when we let him. In turn, Satan can do so much evil when we let him, but he often does so by hiding behind devastating doctrines.

No wonder Christianity sees the history of humanity as a perpetual, cosmic warfare between God and Lucifer, between good and evil, between the Light of God and the darkness of evil, between sound ideologies and corrupted ideologies, between God calling us to be like his image and Satan enticing us to be our own image, which is Satan's image. It is only the religious "eye" that sees all of history in this light: as a cosmic and constant warfare between God and Satan, waged everywhere and daily. It "sees" how life is more of a battleground than a playground. It "sees" how the power of evil and the darkness of Satan can spellbind and enslave the souls, minds, and wills of millions, by infecting them with false doctrines. This explains how we can sell our souls by following "orders" that stem from sources far beyond their own resources. Only religious people are able to see this dimension in history that historians often miss.

Only when we reject the destructive doctrines explained above can we prevent the original beauty of the soul, the mind, and the will from being tarnished any further. Only then can we begin to restore their original beauty that was once given to us in the image and likeness of God. Obviously, this also requires discipline. As the Letter to the Hebrews (12: 11) put it, "For the moment all discipline seems painful rather than pleasant; later it yields the peaceful fruit of righteousness to those who have been trained by it." The question is: do we allow God to enlighten our darkened minds and strengthen our corrupted wills, or do we let Satan take over? Indeed, they can steer us in the right direction.

CHAPTER 7

The Future of Body and Soul after Death

The painful word is out: death. Undoubtedly, facing death is a dreadful experience. This explains why cancer, or any other life-threatening disease for that matter, is perceived by so many as a death sentence. The comedian Woody Allen probably said what most of us think when he jokingly said, "I am not afraid of death, but I don't want to be around when it happens." No wonder, the thought of death scares most people to death. Death is literally taken as the final execution by most people.

But is death really a final execution? That could only be true if a person were merely a body in the process of aging and deteriorating. But what if a person is "both body and soul"? As we discussed earlier, the body may perish but the soul may not. People of faith know their souls do not die when their bodies die. They believe that their death sentence is not a final execution, but instead may even bring them closer to God. Since God is more than a human being, not less, God must at least be a Person, albeit a Person with much more power, much more knowledge, much more love—actually all of these to an infinite degree and perfection. If that's who God is, then we probably want to come as close to God, our origin, as possible—even more so after death. If that is true, then death is not a dead-end but a journey home, back to the Father who created us, to see Him face to face.

Faith tells us that when it's our time, we have to leave our earthly homes in the hope of going to our eternal home. Pope John Paul II's final words in this life were, "Let me go to the house of the Father." He wanted to go back home—to the home that God invites all of us to go to.

The Journey Back Home

S: Despite all the nice things you have said about the beauty of soul, mind, and will, I wonder whether there is really a future for them after death. Can they really surpass death? In other words, is there something real to what you call "a journey back home," or is that only wishful thinking?

R: Earlier I spoke of natural truths versus supernatural truths. Well, the fact that all humans are mortal is a natural truth about their bodies, but the fact that they are also immortal is a supernatural truth about their souls. Human beings do not simply cease to exist when they die, because they are made in God's image and their souls are immortal. The "dust of the earth" is where our bodies come from, but "home" is where our souls come from. The notion of a supernatural truth beyond what meets the eye or is discovered by the scientific method may be hard to accept for many nowadays, but this in itself does not make that notion invalid.

We know home already from our lives on earth: it is the place where we came to know life and learned how to act in life. Not surprisingly, many songs have been written with the magic word "home" in it: *Sweet Home Alabama*, *Take Me Home*; *Home Sweet Home*; *Take Me Home Tonight*; *Mama, I'm Coming Home*; *Homeward Bound*. There is apparently a universal appeal to the word "home." As Chesterton put it, home is "The place where babies are born, where men die, where the drama of mortal life is acted, is not an office or a shop or a bureau. It is something much smaller in size and much larger in scope." Yet often home is "too large and too close to be seen." No matter how we look at it, that kind of earthly home is only a temporary place for us to live in. This leaves the question open as to whether there is an eternal home as well? Is there still a "home" for us when we die and leave that temporary place on earth that we used to call home? Is "home" also meant to be a foreshadowing of heaven? Is there a "journey back home"? Or is "home" gone forever after death?

I think it is hard to claim that a "journey back home" is impossible. The irony is, though, that we cannot prove there is life-after-death, but neither can anyone prove there is *not*. It's more like a wager: if there is nothing after death, we will never know; but if there is indeed an afterlife, that will be the nicest surprise we have ever experienced. True, no one knows for sure what is behind that closed door of death until we have opened that door and have stepped through it.

Therefore, I ask you: should we really take the risk that life-after-death might be just an illusion? If the idea of an afterlife is true, it will be a gift from Heaven—but if it is *not* true, we will never find out. This is what Blaise Pascal called his "Wager." We could describe this wager as follows: if God does not exist, it does not matter how you wager, for there is nothing to win and nothing to lose after death. But if God does exist, then your only chance of winning eternal happiness is to believe, and your only chance of losing it is to refuse to believe. In other words, not-believing is a terrible bet, for it gives you no chance of winning. As Pascal put it, "If you gain, you gain all. If you lose, you lose nothing."

Obviously, the Wager is essentially a form of gambling for sceptics and non-believers. If you are a religious believer, on the other hand, you have no reason to wager. If you firmly believe in God, instead of just wondering whether He exists, then you know God will not abandon you and will eventually invite you back home, for that's why He created you. If death is indeed not a dead-end but a journey-back-home, then the people left behind do not really say "a final goodbye" to their dear ones, but it is more of "a final farewell"—wishing them to "fare well" on their journey back home. Seen in this light, death becomes even more momentous. It is not just the end of a life on earth but the beginning of a new life in Heaven. "Life is short," Cardinal John Henry Newman said. "Death is certain; and the world to come is everlasting." This makes the ultimate much more important than the immediate.

This has quite some consequences for the way we deal with death and with those who are dying. If the end-of-life, as it's usually called, were really the end, closure, and termination of life, then we could finalize the process any way we want—even through mercy-killing, doctor-assisted suicide, active euthanasia, or doctor-prescribed death, as we discussed earlier. But that cannot be true or right. In all these forms, ending life intentionally amounts to suicide—that is, ending your life on your own terms or choosing to die on your own watch.

However, the end of life is not up to us to decide. If life is something we received from God, then we cannot just give it back when we feel ready. The killing of a human being is morally unacceptable, no matter whether it is your own body, someone else's body, or your patient's body. Life is a gift from our Creator which we did not start on our own and therefore cannot end on our own. Think of a gift you received on Christmas or on your birthday; you cannot give it back to the person who gave it to you as a gift. True, in stores, you can return gifts, but that's because the store did not give the gift to you. Well, a gift from God is not up to us to accept or reject. There is no "return desk" for life.

S: But how could we ever know anything about this journey back home? We have never been there before, so we have no experiences to go by. End of the story, I would say as a scientist!

R: You are right, the "journey back home," after death, is not a journey to a place where we had been before. "Back home" is going back to the place where we were meant to be from the very beginning: Paradise—a place that Adam and Eve, including all their descendants, were driven away from. It is a place we never really knew, just like Monarch butterflies fly to a place they have never seen before, and yet they know where to go, following some kind of built-in compass and drive. They go to a place where their ancestors came from, yet unknown to themselves.

Religious people also know where to go when they die—back to the place their first ancestors came from. However, that does not mean we have automatic access to the Paradise from which our first ancestors were expelled after the Fall. Being descendants of Adam and Eve does not entitle us to a guaranteed entrance—we lost that guarantee. Not only did Heaven have to be re-opened for us by Jesus' Crucifixion, but also do we have to gain our original birth rights back again.

Each time I attend a funeral, I marvel at how easily the deceased person is promoted and elevated to Heaven. Our humanistic culture, dominated by the slogan "me, myself, and I," tends to believe that God *owes* us a place in Heaven. One time, a troublemaker asked Archbishop Fulton Sheen a question about someone who had died. The bishop replied, "I will ask him when I get to heaven." The guy replied, "What if he isn't in heaven?" The bishop replied, "Well then you ask him." His response made clear we do not automatically end up in Heaven to meet everyone we had known before on earth.

Yet, if we go by obituaries and funeral speeches, all of us seem to end up in Heaven. According to gravestones and prayer cards for the deceased, we all seem to automatically "rest in peace" after death. If the deceased had incorrigible habits of sin, we are told not to make them feel uncomfortable by publicly condemning the sins they were prone to. Instead, let's tell them that God is merciful, forgiving, and tolerant. Let's also remind them that all sins can be instantaneously wiped away in the confessional or on a good deathbed—or even without those. Above all, let's tell them that you don't have to get an A-plus in sanctity, as a D-minus will do just fine too. Heaven is often seen as something we all receive for "just showing up." Well, the bottom line of all this kind of reasoning is that everyone is supposed to go "back home" where all of us really belong and deserve to be. It has become one of our extended "birth rights."

Unfortunately, there is much confusion about that part of the "back home" story. It is indeed quite common for people to say at a funeral that the deceased person will live on in their memories forever, especially after a heroic battle with a disease they had. But the problem is that memories do not last more than a few decades. After that short period of time, the memory of the deceased will be gone, erased from the face of the earth. Memories of your dear ones are priceless, but if that is all there is left of them, then such memories are useless. Yet, it is common funeral talk to say, "We will never forget you," but as soon as those saying this happen to die too, the memory of "you" dies with them.

Memory is a very fleeting phenomenon—especially in the light of eternity. As the Bible puts it (Sirach 44:8-9), "There are some of them who have left a name, so that men declare their praise. And there are some who have no memorial, who have perished as though they had not lived." How true that is: unless you wrote a famous, timeless book or unless your name is connected to an earth-shaking discovery or invention, you will be forgotten as soon as you have died, or perhaps even sooner. Pharaohs knew that—that's why they built their pyramids. To be remembered by other people lasts only as long as those other people and their memories last. In contrast, the only thing that lasts forever is to be remembered by God.

Unlike the pharaohs, we do not have to build a pyramid for ourselves, for after death we live no longer on planet earth but are meant to be on our way to a life-after-death. "We are here on earth," as the old Baltimore Catechism used to put it, "in order to know and to love God, and to serve

Him in this world, and to someday go to heaven." Do not take Heaven as a smokescreen for the hard life on earth. As a matter of fact, Heaven was first, earth came later—that's a religious fact. Earth was made by God in Heaven and can only exist because there is in fact a Heaven, for without Heaven there would be no earth.

Unfortunately, there is always also a dark side to the "back home" story. Some may never make it to their final destination where they were "meant" to be. Do we have to hear that part as well? Yes, we do, otherwise we hear only a half-truth, and half-truths are lies. The other half of the truth is that the new life after death may not necessarily be in Heaven. Even the famous R.I.P. expression "*requiescat in pace*" on graves and prayer cards is often misunderstood. It is not a command like "Rest in peace," nor a phony promise like "He or she will rest in peace," but only a wish or prayer like "*May* he or she rest in peace." We can only hope and pray that peace will be given to the dead, but it is not a guarantee. The decision is in God's hands.

To ignore this uncomfortable part of the truth sounds to me more like one of those advertisements trying to sell us snake oil—something that promises everyone much more than what it can deliver: a perfect product, without any faults or side-effects, for a low discount price, and with a life-long warrantee. We should know better. There is this joke of a man who had done awful things in life, but when he died, his brother promised the pastor a generous financial kickback if he would mention in his eulogy that the deceased man had actually been a saint. The pastor promised he would do so in exchange for the kickback. Well, when the time of the eulogy came, he gave a short and very honest biography of the deceased person, and then ended with the words, "But compared to his brother, he was a saint." Somehow, we are all declared saints at the moment we die, no matter how we had lived our lives.

S: You are leaving one other part of the truth out: doesn't the Bible tell us that God is a God of Mercy? You seem to ignore, or even deny, this fundamental truth of Christianity. I would say, thanks to God's mercy, we should all end up in heaven!

R: You only focus on one part of what the Bible proclaims: the pleasant truth of God's mercy. But you are ignoring the other part: the hard truth of God's judgment. We will not be judged by our feelings but by our deeds—all of them, that is, by our virtues and vices, our good deeds and our sins. Earlier we made a distinction between natural truths and supernatural truths. What we have here is a supernatural truth—a truth that the Bible is very definite about. Because God is not only a God of Mercy but also a God of Justice, there will be a final judgment for each one of us. During this final judgment, Jesus will judge us according to our deeds (Mt. 25:31-34):

When the Son of man comes in his glory, and all the angels with him, then he will sit on his glorious throne. Before him will be gathered all the nations, and he will separate them one from another as a shepherd separates the sheep from the goats, and he will place the sheep at his right hand, but the goats at the left. Then the King will say to those at his right hand, 'Come, O blessed of my Father, inherit the kingdom prepared for you from the foundation of the world.'

This separation, taking place at the end times, between the ones on the right and those on the left reminds us of the warfare we all have gone through here on earth. There are the two sides again: the ones taking sides with the good forces that come from God versus the ones taking sides with the evil forces that come from Satan. In all of this, God reveals himself not only as a God of *Love* but also as a God of *Justice*. God is both Love and Justice; He is both loving and just. In other words, when God is just, He is also loving, and when He is loving, He is also just—similar to the way parents should deal with their children.

However, there is a certain tension between God's Love and God's Justice when it comes to God's relationship with sinful people. That's where the concept of *mercy* comes in. But mercy-talk could easily obscure that we must first acknowledge that we can do what is wrong. Yet, modern minds seem to ignore that part of God. As the late Fr. James Schall, S.J. put it, "Mercy without judgment bypasses free will." Because we are free human beings, we will be held accountable for our choices in life. Depending on *our* choices, God makes *His* choices. That's where the voluntary part of the human soul comes in, enabling us to have intentions and to act freely.

Obviously, the possibility of moral evil is also a consequence of this human freedom. How could God ever give us freedom without accepting its consequences up to the point of us freely choosing the wrong outcome? We have inherited a nature which is free indeed, but it is weakened and blinded by our own sins and by the sins of our ancestors, right back to the first ones. The Original Sin did not take away the free will. As St. Augustine puts it, "He who created you without your cooperation, will not save you without your cooperation." In other words, God wants our cooperation. We need to *allow* Him to enlighten our darkened minds and to strengthen our corrupted wills. That means every gift God gives us we can refuse. If Adam and Eve, who possessed grace and a perfectly intact nature, could freely sin, how much more so is it possible for all their descendants to sin. If Adam and Eve could fall from grace, surely all of us can fall from grace as well.

Human freedom is always a two-way street, leading us either *toward* God or *away* from God. Consequently, when God gave us a free will, He could not give us freedom without the potentiality for sin and doing evil. This means evil can be, and must be, a very real consequence of human freedom. God can do so much good when we let Him. And Satan can do so much evil when we let him. Sin is the very *worst* there is, since it is the contrary of God who is the very *best* there is.

That is the point where the possibility of either Heaven or hell comes in, making hell as real as Heaven. God's Love and Mercy call many to Heaven, but God's Justice may also have to send some into hell. However, God does not cast persons into hell against their will. God does not lock us up in hell, but we ourselves may do so. No one wants hell to exist; no one wants evil to exist. But hell is just evil eternalized. If there is evil and if there is eternity, then there can be hell—it is a consequence of our own free choice. C. S. Lewis called hell "the greatest monument to human freedom." He also says, "There are only two kinds of people—those who say to God, 'Thy will be done,' or those to whom God says in the end, 'Thy will be done.'"

To put this in a nutshell: if there is life after death, then there must be a Heaven: if there is no Heaven, then there is no hell; if there is no hell, then there is no sin; if there is no sin, then there is no judgement, and if there is no judgment then evil must be as good as good is evil. If, however, we do accept there is good and evil, then there must somehow be a Heaven where good is rewarded as well as a hell where evil is punished. In the words of the Catechism (1039, 1040):

In the presence of Christ, who is Truth itself, the truth of each man's relationship with God will be laid bare. The Last Judgment will reveal even to its furthest consequences the good each person has done or failed to do during his earthly life.

Fortunately, during the Last Judgment, we are not judged by fellow people, but by God. Human beings are just not good or fair enough in judging themselves or others. Therefore, we will be in for many surprises when we find out that the ones we expected to be in Heaven are not there, and the ones we did not expect to find in Heaven do turn out to be there.

S: You have said some hard words here. You probably consider them hard truths. But the fact of the matter is that they create quite some obstacles for all of us on our "journey back home."

R: Indeed, "back home" is good news and bad news at the same time. It is bad news for those who have never been willing to acknowledge their sins. The best we can do is hoping and praying that God may forgive them. The last sentence of the *Hail Mary* prayer cannot be repeated enough: "Pray for us sinners, now and in the hour of our death."

However, there is also good news. The good news is that "back home" is a place where all our tears will be wiped away—tears from broken souls, broken hearts, broken families, broken generations, and a broken world. As the Book of Revelation (21:4) says about God: "he will wipe away every tear from their eyes, and death shall be no more, neither shall there be mourning nor crying nor pain any more, for the former things have passed away."

Apparently, that is good news for all the people who had so many tears and sufferings—and who hasn't? There is this story of a person who had suffered much and had entered a heaven where he found no one who had been broken in life. Not surprisingly, this person could never feel "at home" there. Heaven is not a place for people who have never sinned, neither is it a place for people who were never broken. There is no space in Heaven for people without scars. Even Jesus, after His Resurrection, still carried the scars of His Crucifixion (John 20:15).

Most of us are familiar with the parable of the two sons. It can be found in Matthew 21:28-32. It is the story of a man with two sons who told them to go work in his vineyard. The first son refused, but later obeyed and went. The second son initially expressed obedience, but ultimately disobeyed and refused to work in the vineyard. The son who eventually did the will of his father was the first son because he ended up obeying his father's wish. Jesus then likens the first son to tax collectors and prostitutes—the outcasts in Jewish society—because they believed John the Baptist and accepted his message of "the way of righteousness" (21:32), despite their initial disobedience to the Law.

But some have suggested there should have been at least a third son in the story—the son who left home and never came back. Authors such as André Gide (in his book *The*

Return of the Prodigal Son) and Harold Pinter (in his play *The Homecoming*) have proposed this third possibility. It does seem to make sense. Is it not the reality of life? We all know fallen-away Catholics who never came back to their Catholic Faith before they died, not even at their death bed. They could make for the third son who left home and never came back.

Did Jesus forget that third son, that third possibility? It's hard to believe He did. He knew there are only two possibilities: you stay with the Father, or you walk away from Him but then come back later. There is no way you can stay away forever. Eventually, we will all have to face the Father—either burdened with resentment or relieved with repentance. We all go ultimately back to the Father. But whether He will let us enter His eternal home, Heaven, is an entirely different question. We could end up in the "pigs' stable" of the Father's House. In other words, that third son never existed.

Most of us want to be welcomed back home—who would not? People have a natural yearning to be waited for and welcomed back, both in this life and in the life to come. People can keep their sanity as long as there is at least one person who is waiting for them. And the good news is this: that one person is God. But it certainly helps when other people are there for you too, both before, during, and after you die. But the only thing that ultimately counts is God's waiting for us.

S: You make it look like we end up, after death, either in heaven or in hell. I was always told Catholics also acknowledge an in-between state called *purgatory*. Is that true?

R: I am glad you are bringing this up, otherwise we would end up with a very uncompromising black-and-white picture—either eternal salvation or eternal damnation. Fortunately, says the Catholic Church, there is also something in between Heaven and hell, a kind of "middle state" called *Purgatory*. It is a place or condition of temporal punishment for those who depart this life in God's grace. The Catechism (1030) says about Purgatory: "All who die in God's grace and friendship, but still imperfectly purified... undergo purification, so as to achieve the holiness necessary to enter the joy of heaven." The idea of purgatory is not exclusively Catholic, though. Current Protestants, while they avoid the word "purgatory," now often speak of "the middle state." Orthodox Christians speak in terms of a "theosis," a journey of transformation.

What does this mean? I think this is best explained by an example the late and legendary Mother Angelica of EWTN once used: if a prostitute had a profound conversion and decides to enter Mother Angelica's convent, a one-day transition would definitely be too short a period for such a person to make this transition—actually a massive shock—in spite of all her good intentions. Indeed, for most of us, the transition from a life on earth to a life in Heaven would be equally dramatic, actually so shocking that we would need some extra preparation time, as nothing unclean can enter the presence of God, according to the Book of Revelation (21:27). We need some kind of "re-education" before we can "graduate" to Heaven.

As the late Fr. Benedict Groeschel put it, "Purgatory is not a temporary hell, but a preliminary heaven." It is a place or state where human imperfection is corrected in the "fire of *purification*" before we can enter God's Heaven where nothing unclean shall enter. The idea of purification is counter to the cheap optimism that prevails nowadays in the minds of many, holding that the life of practically everybody automatically ends up in a state of bliss. However, the truth is that not everybody's life ends up in a state of ecstasy, as we discussed already. Since we all have stained souls—no one is without sin (Rom. 9:23)—the only way to remove such stains is repentance and remorse and purification. Cardinal Newman once described a soul coming to the judgment seat of Christ, begging his guardian angel to take him away to Purgatory since he is keenly aware of his need for purification.

True, there is also God's infinite mercy. But mercy does not wipe out all evil, only the evil that we ask forgiveness for. God's mercy is boundless, but to access it requires repentance. "Lord, have mercy" does not translate into "God, go easy on me." Mercy does not mean to be "soft on crime."

God's mercy and His justice go hand in hand. St. John (1 John 1:8-9) makes this very clear to us when he says:

If we say we have no sin, we deceive ourselves, and the truth is not in us. 9 If we confess our sins, he is faithful and just, and will forgive our sins and cleanse us from all unrighteousness.

Hence, St. Augustine advises us, "After sin, hope for mercy; before sin, fear justice." St. Ambrose said about the deceased, "We have loved them during life; let us not abandon them until we have conducted them by our prayers into the house of the Lord." Such prayers would make no sense if these early Christians thought the souls of the dead were already either in hell or in Heaven, for they could no longer benefit from prayers. Thank God, there is also a place where we can be purified from our sins. In purgatory, we can see in all clarity who we were and where we came from before we can enter the eternal glory of God.

A Transition through the Tunnel of Death

S: Is there a way to picture the transition from mind and soul to the eternity of heaven? Or is this return "back home" completely beyond human imagination? The scientist in me needs more information.

R: That is indeed an important question. What could the way-back-home look like? Did we ever personally experience anything that comes close to it? I think we did when we were born into this world. I would be surprised, though, if anyone has still memories of how they came into this world at birth. Yet, based on biology and psychology, we do have an idea of what this experience must have been like.

When we were born, we were "expelled" from a dark womb into a world completely unknown to us. Even when there are no complications at the time of birth, various dramatic changes must take place for every baby at the moment of birth. The placental circulation must be cut off; the temporary connection between the pulmonary artery and the aorta needs to be closed; the temporary openings in the wall between the left and right heart chambers must be closed; the lungs will be inflated for the first time; and blood will be forced into the pulmonary circulation system. We all have gone through this process successfully—and hopefully unharmed. We were never asked whether we wanted to be born—we just had to surrender to this transition. That is how we entered our "home" here on earth.

If this is true—and there is not much reason to doubt it is, as it amounts to a natural truth—then not only is death a necessary part of life, but it could also be a necessary *stage* in life—a stage of transition into another kind of life perhaps, not an ending but a transformation. Seen in this light, what we call the last stage may very well turn out to be the next-to-last stage, when seen in a wider context as a supernatural truth.

The surprising fact is that, when we came into this world, we were forced to leave the protective shelter of our mother's womb. At that moment, rather abruptly, life in the womb was over. It must have been a shocking, almost traumatic, experience when we were pushed out of her womb, forced to go through the tunnel of her birth canal, and pressured to breathe on our own—a physical and emotional shock, for sure. At that moment, we were certainly entering an entirely different and unknown world. We had to trust that we could live there and that there were hands welcoming us. We were fortunate enough to find out that there was life beyond life in the womb.

Unfortunately, though, we must also add that some did not find "hands" welcoming them outside the mother's womb. They had to experience at birth that their father, or even their mother, had forsaken them—which creates a lifelong brokenness of heart. Broken families create broken hearts as part of the Original Sin. But fortunately, God has taken on this brokenness. For even if there were no hands waiting for us at birth, there still will be "hands" waiting for us at death. As Psalm 27:10 tells us: "Even if my father and mother forsake me, the Lord will take me in." And in Isaiah (49:15-16), God touches our broken hearts:

Can a mother forget her infant, be without tenderness for the child of her womb? Even should she forget, I will never forget you. See, upon the palms of my hands I have engraved you.

Nevertheless, the fact remains that when you die, you leave behind "life"—your own life. If death is really and truly death, you will be destroyed forever—never again to be "you," never again to think or to feel, never again to laugh or to love. But if there is something beyond biological death, then there must be another dimension of life that we may have overlooked or misunderstood—an immaterial dimension not limited by space and time. True, if we are just biological machines, we certainly die—and that is it. But if we are more than biological machines—bodies with a beautiful soul, including mind and will—then biological death may not be the end.

True, death is certainly a fearful experience, as it is saying farewell to all that was dear to us, as well as to all that was not so dear to us. When we are being pushed through the "birth canal of death," we leave behind the comforting world we had become so accustomed to. But if the analogies between birth and death hold true, we should not be so fearful about this transition, for just as this life surpassed existence within the womb, so too could the next life surpass all we have experienced here on earth—with "hands" waiting to receive us at the other end.

Why would anyone choose to remain unborn after having experienced even some of the riches of life? Something similar may hold for embarking on a journey beyond death.

As a matter of fact, we have no choice but to embark on this journey beyond death. We may feel like Monarch butterflies who "know" when to migrate to a land they have never been to, along a route they have never followed before. In a similar way, the human soul also "knows" there is a time to "migrate" to where its eternal "home" is. It returns to a place it has never been to before. Yet, we were born with an innate desire to look beyond life on earth: to the grass on the other side of the mountain, to the land on the other side of the ocean, to the stars beyond the stars that we can see—and also, which is much more important, to a life on the other side of the tunnel of death.

C. S. Lewis placed all of this in a wider context when he wrote:

Creatures are not born with desires unless satisfaction for those desires exists. A baby feels hunger: well, there is such a thing as food. A duckling wants to swim: well, there is such a thing as water. Men feel sexual desire: well, there is such a thing as sex. If I find in myself a desire which no experience in this world can satisfy, the most probable explanation is that I was made for another world.

The fact that our heart yearns for something Earth can't supply is proof that Heaven must be our home.

S: So far, you only talked about the soul and how the soul goes through the tunnel of death. But what about the body? Isn't the body left behind on earth to decay and return to dust? But then I ask you, how can this be, for this would violate the unity of body and soul you talked about earlier? On the other hand, claiming that the soul can exist without the body seems to be an inconsistency in Aquinian philosophy and yours.

R: You are right, that is what it looks like, an inconsistency. Even St. Thomas Aquinas himself readily admits this: "It is clear that the soul is united to the body by nature: because by its essence it is the form of the body. Therefore, it is contrary to the nature of the soul to be deprived of the body." However, that does not mean that the soul cannot exist on its own after death. Aquinas regarded the soul as a "substantial form," making it independent of the body in its existence—in the same way as the human intellect does not require any bodily organ for its operation. It is therefore of such a nature that it can subsist on its own, even after the dissolution of the body of which it is the "form."

This idea of an independent soul, despite its dependence on matter, may balance on the verge of contradiction, but it is not as strange as it sounds. Think of the analogy of language: We can only think with and through language, yet our thinking surpasses language and is not absorbed by it, despite our dependence on it. In a similar way, although an incarnated soul is fully tied to matter, the very "act of incarnation" indicates there *is* something that "enters" matter. Nowadays we know even better than Aquinas did that the body keeps constantly changing while the soul persists. Therefore, one could argue that if the soul persists while the body keeps changing, then the soul may also persist when the body is corrupted by death.

Nonetheless, although human reason tells us man is mortal—even if his soul is immortal—it is contrary to the nature of the soul to be deprived of the body. Since a person is a unity of body and soul, death separates the body from the soul, so it separates us from ourselves. A soul stripped completely of its body is literally a no-body. It is not a complete substance like it was for Descartes, but an incomplete substance (think of a heart that temporarily subsists on its own after taken from a donor). When you die, your soul carries on, but you as a person do not. However, the survival of your soul makes it possible for you as a person to live again, God willing. Consequently, says Aquinas, when separated from the body, the soul is in a "violent" state, and the desire of the immortal soul is to be joined again with its own body. With this insight, Aquinas anticipated and foreshadowed the Resurrection of the body in his philosophy.

Long before St. Thomas Aquinas had argued all of this, the Bible had already declared the Resurrection of the Body a supernatural truth. It is an important issue in Catholic Faith. For that reason, the core creeds of the Church—the

Apostles' Creed, the Nicene Creed, and the Athanasian Creed—mention explicitly "the resurrection of the body." This has been the historic teaching of the Christian faith going back to its very beginning. In other words, body and soul must be united again in Heaven according to the unity they had on earth. This does not happen immediately when we die, but it will happen during the resurrection of the body at the end of time.

The Apostle Paul teaches that all Christians, even those whose bodies had "perished," would be given new, resurrected bodies (1 Cor. 15). This means that after death, one's departed soul will eventually be restored—resurrected, that is—to a bodily life in Heaven. As the Catechism of the Catholic Church (#997) puts it: "In death, the separation of the soul from the body, the human body decays and the soul goes to meet God, while awaiting its reunion with its glorified body. God, in his almighty power, will definitively grant incorruptible life to our bodies by reuniting them with our souls, through the power of Jesus' Resurrection."

The question then is: *when* will this happen? Common biblical expressions for this event are "on the last day," "at the end of time," "when Jesus returns back to earth," "at the second coming"—apparently, not immediately after death. In other words, there seems to be a particular judgment, immediately after a person's death, as well as a general judgment, at the end of time for everyone. Each individual person is judged by God immediately after his or her death.

But the Bible is very emphatic in asserting that there will also be a general judgment of all mankind at the end of the world. Whereas at the *particular* judgment, we will know instantly all the good and the bad we have done, at the *general* judgment will we see what effect our moral acts had on others around us and on other generations after us. Whereas the particular judgment addresses all of us as individuals, the general judgment will address all of us collectively as members of one large family, the human race.

Just as the time of reckoning arrives at last for the individual at the end of life, so it does for the world as a whole at the end of time. Why the latter? There is such a huge amount of injustice in human history which must be set straight by God's Justice. Think of this: many have died or will die without ever having been welcomed into life, without ever having received love, comfort, or even justice from their fellow people. What about all those victims of genocide, gas chambers, torture dungeons, terrorism, wars, drugs? What about all those people who cannot be called back to life to receive a bit more warmth and love here? What about those who were neglected by their spouses, their parents, or their children? This explains why, in addition to a particular judgement, there is a need for a general judgment, the Last Judgment. Put differently, there are many "debit accounts" that still need to be settled. Without a final judgment, those accounts would remain unsettled.

The next question is: *what* will happen then, at the end of time? The Bible says that when Jesus returns to earth, he will physically raise all those who have died, giving back to them the bodies they had lost at death. How can this be? How can we have a new body and yet the same body as we had on earth? Let's not forget that even on earth we had the material for many bodies, because every cell is constantly being torn down and rebuilt during our lifetime on earth. But if that is the case, why then could we not have the same body as we had before death?

It is not a new body that will be given to us; it will be our own present bodies given back to us. After death, God can recall and rebuild the material of the body, but without imperfections. The bodies that all human beings will possess forever will be their own bodies that they had on earth. But there is one important difference: in the words of St. Paul, "It is sown a physical body, it is raised a spiritual body. If there is a physical body, there is also a spiritual body" (1 Cor. 15:44).

Some Christians today may think that the resurrection of the body would be difficult for those who have lost parts of their body during life on earth or have lost their physical body after cremation on earth. However, the Bible assures us that God can raise the bodies of all who have died, no matter when or how they died. God can restore their bodies, just as when He created them, but now in a glorified form. As St. Paul (1 Cor. 15:53) puts it, "this perishable nature must put on the imperishable, and this mortal nature must put on immortality."

One thing must be made clear, though, before we close this part of the discussion. Resurrection of the body is not something like *reincarnation.* According to a 2018 Pew survey, 29% of Christians in the U.S. hold the New Age belief of reincarnation—the belief that when one's body dies, one's soul lives on in a new and different body, unrelated to the first. However, the idea of reincarnation is based on dualism and Gnosticism, and therefore definitely not orthodox Catholic. It ignores the strong unity between body and soul, with the soul expressing itself through the body, which makes the body an essential part of who we are.

What is Heaven Like?

S: This makes me wonder how we could ever talk meaningfully about life on the other side of death. No one, other than Jesus Christ, has come back to earth to tell us what heaven is like, for being in heaven means not being on earth anymore.

R: You are right, we have a challenging issue here. How can we ever capture something from Heaven while we are still on earth? St. Paul gives us a glimpse of the answer: "For now we see in a mirror dimly, but then face to face. Now I know in part; then I shall understand fully, even as I have been fully understood" (1 Cor. 13:12). We all know here on earth that we are finite beings with a finite mind—if you do not know this, think again, for it is a natural truth. But somehow our human mind, finite as it is, is able to catch a glimpse of the Infinite. To use a poor analogy: if I can count from one to ten, or from one to thousand, then I can also count from one to infinity, at least in theory. Somehow, we have the inborn capacity to reach out to infinity with our minds.

Applied to ourselves, the human mind has the capacity to reach beyond itself. Earlier we called this self-transcendence—referring to the transcendence of something, or rather Someone, more than our own selves. We could apply this also to our everyday experience of imperfections in justice, love, goodness, and beauty—which indicates an innate sense of perfection that brings us to apprehend the possibility of encountering perfect justice, perfect love, etc. It is a way of reaching out to infinity. The only place where this can really come true is in Heaven. However, thanks to the beauty of the human mind, we do have some access to what transcends us. We are able to catch a glimpse of what is "above" and "beyond us"—which might include a vague indication of what Heaven is like.

It needs to be stated first that life in Heaven cannot be a pure continuation of life the way it was on earth. If you were married on earth, for instance, that does not mean you will be married in Heaven, too. Jesus himself warns us for such simplicity. In Lk. 20:34-38, he says: "The sons of this age marry and are given in marriage; but those who are accounted worthy to attain to that age and to the resurrection from the dead neither marry nor are given in marriage." And in Mt. 22:30, Jesus says, "For in the resurrection they neither marry nor are given in marriage, but are like angels in heaven." If anyone would ever tell you that they could not be happy in Heaven without their spouse, one might as well say that if they ever find themselves in that situation, one may be sure the situation they are in is not Heaven. Heaven is being with God rather than with one's spouse.

Nevertheless, most people have a rather primitive, childlike idea about Heaven. The Muslim heaven, for instance, is a man's haven where each man is rewarded with seventy-two beautiful, high-bosomed virgins, plentiful food, slaves galore to attend to every whim and wish (Qur'an 78:31-34). Apparently, the bliss there comes from base sensual pleasures—food, drink, and sex—which is in essence a continuation of the way it could have been on earth.

Even in the eyes of some Christians, their own Christian Heaven is not much different. They portray Heaven as a place of eternal picnics and "lions laying down with lambs." But we should not forget that lions, lambs, and picnics get boring after a while. Or they picture Heaven as an endless repetition of alleluias or the monotonous fingering of harps. Even that can get boring! Or they "eternalize" what they liked doing on earth—playing piano, playing golf, puttering in the backyard, and so on—but there are no pianos, golf courses, and backyards in Heaven. C. S. Lewis could not have said it better: "It is quite useless knocking at the door of heaven for earthly comfort. It's not the sort of comfort they supply there." No wonder, in one of his books, Cardinal Ratzinger warned us against depicting Heaven as an extension of this life prettied up with idyllic depictions. It is not like a luxurious resort or a luxurious retirement home.

If that is not it, what then is Heaven really like? To put it in a nutshell, the answer is this: the bliss of Heaven comes from the pure joy of being in God's presence. As St. Augustine said about God, "our hearts will never be at rest until they rest in You." This "rest" is more a *state* of being than a *place* somewhere in the clouds. As the Catechism (2794) says about the line "Our Father who art in heaven" at the beginning of the *Our Father*,

This biblical expression does not mean a place ("space"), but a way of being; it does not mean that God is distant, but majestic. Our Father is not "elsewhere": he transcends everything we can conceive of his holiness. It is precisely because he is thrice holy that he is so close to the humble and contrite heart.

As hell is a state of eternal damnation, Heaven is a state of eternal salvation. The language of "place" is, according to Pope John Paul II, inadequate to describe the realities involved, since it is tied to the spatiotemporal order in which this world and all of us exist. Although the Book of Revelation does provide a physical description of Heaven—the New Jerusalem with walls of jasper, gates of pearls, and streets of gold—we must be mindful that such imagery conveys the inexpressible and incomprehensible beauty of Heaven rather than intending to be a literal description. God does not occupy some remote geographic corner of the physical universe, and no MapQuest search can tell us where Heaven is. It is a "state" rather than a "place."

S: Okay, a "state" then, instead of a "place." But that still does not sound as much—not something to look forward to or dream about.

R: I would say it is in fact everything we could ever dream of. We will see God face to face, not with physical eyeballs but with the spiritual "eyes" of the soul. Our physical eyes fail here. Let us not forget that St. Paul actually warns us not even to try conjecturing what heaven will be like: "no eye has seen, nor ear heard, nor the heart of man conceived, what God has prepared for those who love him" (1 Cor. 2:9). Or as St. John puts it, "we shall be like him, for we shall see him as he is" (1 John 3:2). Yet, it is in Heaven that God "will wipe away every tear from their eyes, and death shall be no more, neither shall there be mourning nor crying nor pain any more, for the former things have passed away" (Rev. 21:4). This means that no sickness or any imperfection will be possible in Heaven. This implies much more than you might think, though: in Heaven, questions we always had will be answered. And so much more, according to the Catechism (2040):

We shall know the ultimate meaning of the whole work of creation and of the entire economy of salvation and understand the marvelous ways by which [God's] Providence led everything toward its final end.

As Pope John Paul II pointed out on July 28, 1999:

In the context of Revelation, we know that the "heaven" or "happiness" in which we will find ourselves is neither an abstraction nor a physical place in the clouds, but a living, personal relationship with the Holy Trinity. It is our meeting with the Father which takes place in the risen Christ through the communion of the Holy Spirit.

Archbishop Fulton Sheen had his own way of explaining this,

Heaven is not a place where there is the mere vocal repetition of alleluias or the monotonous fingering of harps. Heaven is a place where we find the fullness of all the fine things we enjoy on this earth. Heaven is a place where we find in its plentitude those things which slake the thirst of hearts, satisfy the hunger of starving minds, and give rest to unrequited love. Heaven is the communion with perfect Life, perfect Truth, and perfect Love.

After stressing that Heaven is not a geographical location because God is outside of time and space, Cardinal Newman goes on:

To go to heaven is to go to God. God is there and God alone: for perfect bliss is there and nothing else, and none can be blessed who is not bathed and hidden and absorbed in the glory of the Divine Nature.

In short, Heaven is our ultimate destiny, back home, where all our brokenness will be healed, and all our tears will be wiped away. Despite humanity's inability to save itself from its brokenness, only God is the one who heals a broken world in Heaven. In the words of the Catechism (1024), "Heaven is the ultimate end and fulfillment of the deepest human longings, the state of supreme, definitive happiness." True and everlasting happiness can only be possible with God and in the presence of God. When we meet God face to face for the first time, we will have something that is a million times more spectacular than any relationship we had on earth.

How does one enter Heaven? As we said before, in order to come to God and be saved, one needs to repent, have faith and hope. No one can be saved who does not want to be. Archbishop Fulton Sheen once said, "Had Christ remained on earth, sight would have taken the place of faith. In heaven, there will be no faith because His followers will see; there will be no hope, because they will possess; but there will be love for love endureth forever!" Although in Heaven there will be no faith, since faith is of things unseen, nor hope, since hope is of things not yet possessed, Heaven is our ultimate hope while we are still on earth. That's why the ultimate is so much more important than the immediate. The Catechism (1818) says about hope:

The virtue of hope responds to the aspiration to happiness which God has placed in the heart of every man; it takes up the hopes that inspire men's activities and purifies them so as to order them to the Kingdom of heaven; it keeps man from discouragement; it sustains him during times of abandonment; it opens up his heart in expectation of eternal beatitude.

Hope is the driving force behind our lives on earth. As C. S. Lewis puts it: "If you read history, you will find that the Christians who did most for the present world were just those who thought most of the next." Obviously, if it were true that there is no God and no Heaven, then there would be no hope for this broken world—with its broken souls, hearts, families, and generations. This broken world needs hope. Therefore, let us end with what St. Teresa of Ávila says about hope:

Hope, O my soul, hope. You know neither the day nor the hour. Watch carefully, for everything passes quickly, even though your impatience makes doubtful what is certain, and turns a very short time into a long one. Dream that the more you struggle, the more you prove the love that you bear your God, and the more you will rejoice one day with your Beloved, in a happiness and rapture that can never end.

Index

About the Author

Gerard M. Verschuuren is a human biologist, specialized in human genetics. He also earned a doctorate in the philosophy of science. He studied and worked at universities in Europe and the United States. Currently semi-retired, he spends most of his time as a writer, speaker, and consultant on the interface of science and religion, faith and reason.

Some of his most recent books are:

- *God and Evolution?—Science Meets Faith.* (Boston, MA: Pauline Books, 2012).

- *The Destiny of the Universe—In Pursuit of the Great Unknown.* (St. Paul, MN: Paragon House, 2014).
- *Five Anti-Catholic Myths—Slavery, Crusades, Inquisition, Galileo, Holocaust.* (Kettering, OH: Angelico Press, 2015).
- *Life's Journey—A Guide from Conception to Growing Up, Growing Old, and Natural Death.* (Kettering, OH: Angelico Press, 2016).
- *Aquinas and Modern Science—A New Synthesis of Faith and Reason.* (Kettering, OH: Angelico Press, 2016).
- *Faith and Reason—The Cradle of Truth.* (St. Louis, MO: En Route Books, 2017).
- *The Myth of an Anti-Science Church—Galileo, Darwin, Teilhard, Hawking, Dawkins.* (Kettering, OH: Angelico Press, 2018).
- *The First Christians—Keeping the Faith in Times of Trouble.* (St. Louis, MO: En Route Books, 2018).
- *The Eclipse of God—Is Religion on the Way out?* (St. Louis, MO: En Route Books, 2018).
- *Forty Anti-Catholic Lies.* (Manchester, NH: Sophia Institute Press, 2018.
- *Broken Hearts in a Broken World.* (St. Louis, MO: En Route Books, 2018).
- *In the Beginning*-How God Made Earth Our Home. (Manchester, NH: Sophia Institute Press, 2019).

- *The Destructive Doctrines of Our Age.* (St. Louis, MO: En Route Books, 2019).
- *Can Science and Religious Live Together*? (St. Louis, MO: En Route Books, 2020).
- *60 Catholics Who Changed the World* (St. Louis, MO: En Route Books, 2020).

For more info:
http://en.wikipedia.org/wiki/Gerard_Verschuuren.

Dr. Verschuuren may be contacted at www.where-do-we-come-from.com.

Made in United States
Cleveland, OH
08 November 2026

47789690R00156